# DR DAVID LEWIS

# Impulse

*Why We Do What We Do
Without Knowing Why We Do It*

BOOKS

Published by Random House Business Books 2014

2 4 6 8 10 9 7 5 3 1

Copyright © David Lewis 2013

David Lewis has asserted his right under the Copyright, Designs
and Patents Act, 1988, to be identified as the author of this work

This book is sold subject to the condition that it shall not, by way of trade or otherwise, be lent,
resold, hired out, or otherwise circulated without the publisher's prior consent in any form of
binding or cover other than that in which it is published and without a similar condition,
including this condition, being imposed on the subsequent purchaser.

First published in Great Britain in 2013 by Random House Business Books
Random House, 20 Vauxhall Bridge Road,
London SW1V 2SA

www.randomhouse.co.uk

Addresses for companies within The Random House Group Limited can be found at:
www.randomhouse.co.uk/offices.htm

The Random House Group Limited Reg. No. 954009

A CIP catalogue record for this book
is available from the British Library

ISBN 9781847946867

The Random House Group Limited supports the Forest Stewardship Council® (FSC®),
the leading international forest-certification organisation. Our books carrying the FSC label
are printed on FSC®-certified paper. FSC is the only forest-certification scheme supported
by the leading environmental organisations, including Greenpeace. Our paper procurement
policy can be found at: www.randomhouse.co.uk/environment

Typeset by Palimpsest Book Production Ltd, Falkirk, Stirlingshire
Printed and bound in Great Britain by CPI Group (UK) Ltd, Croydon CR0 4YY

To S.M. with my love and gratitude.

# Contents

'Instinct leads, intelligence does but follow.' William James, *The Varieties of Religious Experience: A Study in Human Nature*, 1902.

# Acknowledgements

Many people have assisted me in writing this book and I am especially grateful to those members of the Impulse Research Group for their support and invaluable contribution.

My special thanks to Dr John Storey for his medical insights and valuable comments on early drafts of the text. I would like to thank my colleagues at Mindlab International, Joe Hilling MSc, Director of Data Analysis and Duncan Smith, managing director for their assistance. Also my thanks to neuroscientist Charlie Rose for his work on the references and critique of the text. Steven Matthews has, as always, been diligent in reading and commenting on the text. I extend my warmest thanks to Dr Margaret Yufera-Leitch of the University of Calgary, one of the world's foremost experts on obesity whose contribution to the chapter on impulsive eating was significant and important. I am most grateful to Coastguard station chief Don Ellis for his insights into the minds of suicides, knowledge gained by often dangerous experience in recovering the bodies of those who were successful and talking others to safety. My thanks also go to Dr Sean Kelly for his valuable assistance.

Finally, I am most grateful to Nigel Wilcockson and Sophie Lazar, my editors at Random House, for their invaluable and essential contributions to the shaping of the book.

Where necessary, names and personal details in the case histories cited have been changed for reasons of confidentiality and to protect privacy.

# Copyright acknowledgements

I would like to thank Dr Hannah Faye Chua for permission to use two images from 'How we See it: Culturally Different Eye Movement Patterns Over Visual Scenes' from the article by Boland, Chua, and Nisbett (2005). In Rayner, K., Shen, D., Bai, X., Yan, G., (Eds.) *Cognitive and Cultural Influences on Eye Movements*. Tianjin, China: People's Press/Psychology Press, pages 363–378.

Also I am most grateful to Dr Richard Russell, Assistant Professor of Psychology at Gettysburg College, for kindly allowing me to use two photographs from his excellent paper 'A sex difference in facial contrast and its exaggeration by cosmetics', which appeared in the 2009 issue of *Perception*.

Thanks to Meg Bury of the explOratorium in San Francisco for permission to use a photograph of the Ames Room which forms part of their permanent exhibition.

To Dr Russell Swerdlow, co-author of the 2003 *Archives of Neurology* paper: Right orbitofrontal tumor with pedophilia symptom and constructional apraxia sign, for his kindly assistance.

My thanks go to artist Terry Ayling for creating the line and tone illustrations and to Norman Clark for constructing a model of the Ames room used in my research.

The images on page 81 are reproduced by kind permission of: *Figure 1*: The Thinker, 1880-81 (bronze), Rodin, Auguste (1840-1917) / Burrell Collection, Glasgow, Scotland / © Culture and Sport Glasgow (Museums) / The Bridgeman Art Library. *Figure 2*: Discobolus, copy of a Greek original (plaster) (b/w photo), Myron, (fl.c.450 BC) (after) / Museo Nazionale Romano, Rome, Italy / Alinari / The Bridgeman Art Library. The images of female and male models on page 116 are reproduced courtesy of Bigstockphoto. com.

# Introduction

As I write these words, my head is watching me from across the room. Every so often the eyes – a colour match for my own – blink. From time to time the mouth silently opens and closes. Created from a life cast that involved pouring gloopy blue resin all over my head and shoulders, while I breathed through straws stuck up each nostril, it is a perfect replica.

Perfect, that is, save for one rather sobering difference. While I have aged, my second head, created more than a decade ago, remains – like some three-dimensional Dorian Gray in reverse – youthful. It still has a full head of hair – each one individually and painstakingly inserted by hand – while mine is fast receding. Its brow is smooth while mine is furrowed. An even more important difference is that you can lift the top of the head of my doppelgänger and remove the brain. Indeed I often tell people I originally had it made as an attention-grabbing lecture aid for teaching neuropsychology. This, at least, is the explanation I usually offer when asked why on earth I invested so much time and hard-earned cash in such a bizarre acquisition. In truth it is really a justification for what was an irrational and absurd impulse purchase. A purchase which I can only explain away, as is the case with most impulsive acts, by saying it seemed like a good idea at the time!

While the impulse that led me to have a second head created was one of my more expensive, it was by no means my most life-changing. Of these there have, so far, been three. Two altered the course of my life completely while the third, which I describe in Chapter 1, saved it.

My first life-transforming impulse occurred when I was 21 and browsing in a second-hand bookshop on London's Charing Cross Road. Until that epiphany, my aim had been to qualify as a medical doctor. When I was ten my parents, knowing I had a great interest in biology, gave me a proper grown-up laboratory microscope. For the next eight years I spent much of my free time dissecting everything from road kill to sheep's eyes and ox hearts to pig brains obtained from our friendly local butcher. To record my dissections I learned how to take photographs and soon expanded my photographic interests to record local news events, selling the pictures to local newspapers to help finance my science studies. In time I was accepted into medical school and started working to achieve my lifetime's ambition. Eighteen months into my medical studies, wandering aimlessly around that second-hand bookshop, I chanced upon a copy of *People I Have Shot* by James Jarché, one of Fleet Street's earliest press photographers. Jarché's life story fascinated me and rekindled my interest in photography. I spent half the night reading it and, the following morning, hunched over a formaldehyde-soaked limb in the dissecting room, decided on an impulse to give up medicine and study photography instead.

The next afternoon I applied for and was given a place on a three-year photography course at the then Regent Street Polytechnic School of Photography, now part of the University of Westminster. When the course finished I moved first to Paris and then back to London to work as a photojournalist for magazines such as *Paris Match, Stern, Oggi* and *Life.* It was a career I pursued with interest and enjoyment for some years until changing course for the second time. Once again on an impulse that occurred at precisely 4.30 on the afternoon of 23 December 1976. A few weeks earlier, a meeting with a clinical psychologist, whose work was the subject of a magazine assignment, opened my eyes to the fascinating and important discipline of psychology. While drinking a cup of tea in the picture agency's Fleet Street offices I took a spur-of-the-moment decision to return to university and read psychology.

That impulse resulted in a complete change of direction in my life and a career that still fascinates me more than three decades later. It has also led to my long-standing professional interest in the psychology of 'impulsivity', in trying to make sense of those spur-of-the-moment decisions that often change our lives. To understand our behaviour in all its many and varied facets, it is essential to understand impulses, since these (as I shall explain in Chapter 2) comprise the vast majority of our actions. We all like to regard ourselves as rational human beings. To believe we act only after careful reflection and thoughtful deliberation. The fact is, however, that our actions are *mindless* far more often than they are *mindful*: the product not of logic and reason but of habits driven by emotions.

Whenever we are motivated by joy or anger, jealousy or envy, love or lust, compassion or avarice, hatred or a desire for revenge we think, speak and act impulsively. We blurt out indiscretions, rush to judgements, make snap decisions, jump to conclusions, take leaps of faith and trust gut feelings far more than rational analysis. It doesn't matter that we know better! It doesn't matter that we're aware of the remorse we will later feel for our impulsive deeds. At some point self-control simply deserts us and we fall in love with the wrong person, impulsively buy things we don't really need, take reckless financial risks, agree to a second helping of that delicious but waist-expanding chocolate cake, allow 'a couple of drinks' to turn into a steady flow of alcohol until closing time, jeopardise our health through overindulgence or succumb to peer-group pressure and participate in risky pranks. Impulses lie at the root of most personal and social problems ranging from obesity, alcohol and drug abuse, overspending, unwanted pregnancies, smoking, emotional problems, dysfunctional relationships, and school underachievement to a failure in achieving cherished life goals.

## What is an impulse?

In some branches of science 'impulse' has an unambiguous meaning. To the physicist it is 'an indefinitely large force that acts for a very short time that brings about a change of momentum'. As, for example, when we hit a nail with a hammer. To the neurologist it is 'a wave of physical and chemical excitation along a nerve fibre in response to a stimulus'. Surprisingly, after almost a century of study, psychologists have still reached no clear consensus as to precisely what they mean by an impulse or why some people are so much more impulsive than others. One of the earliest attempts at a definition was made in 1890 by the pioneering American psychologist William James. 'Impulses,' he wrote, '[are] ephemeral thoughts usually tied to forceful urges.' Many of the definitions offered by later psychologists were equally woolly and, on occasions, mutually contradictory.[1]

In the following chapter I shall be providing my own definition of an impulse when I explore two distinct ways of thinking that operate within the single brain. For the moment let's just note that all the above definitions, whether from physics, biology, psychology or ordinary speech, express a common idea: that of getting things moving. The word itself comes from the Latin *impulsus*, which is related to *impellere* 'to move', and many impulses are indeed life-changing events. Small wonder, then, that they have fascinated philosophers, theologians and thinkers since the dawn of civilisation.

# The impulse in history

The Greek physician Hippocrates offered one of the first explanations for impulsive behaviour in the fifth century BCE. He based his theory on the belief that a man's personality depends on the balance between the four bodily fluids of black and yellow bile, phlegm and blood. A melancholy man was thought to suffer from an excess of black bile and a phlegmatic individual from too much phlegm. Impulsivity, Hippocrates believed, was due to excessive yellow bile producing a personality whose impulses were caused by anger – what he termed *kakia* or emotional evil. Too much blood in the mix, by contrast, resulted in a sanguine personality whose impulses were caused by *aporia*, which roughly translates as 'perplexity'.

For Christians, and the other Abrahamic religions of Islam and Judaism, impulses presented a challenge that the faithful argued over for centuries. Since they believed God to be perfect, it followed that he would not create anything so imperfect as to exhibit impulsive behaviours. Yet humans often acted in ways that were manifestly impulsive. Their solution to this apparent contradiction was to invent the devil and blame all impulses on him.[2]

The 15th and 16th centuries were Satan's golden age. It was a period during which plunging temperatures during Europe's mini-ice age led to crop failures which, combined with rising populations, left many starving. Between 1683 and 1684, during what became known as the Great Frost, the Thames froze over so completely for two months that oxen were roasted on its centre. Ice nearly a foot thick covered the tidal waters and unbroken ice extending miles off the coasts of the England, France and the Low Countries blocked many harbours and caused severe problems for shipping. For nations afflicted by disease, crime and endless wars these meteorological disasters were seen as omens and portents.

Among the faithful, there was bewilderment that a merciful God could cause such pain and suffering. Around 1486, Jacobus Sprenger and Henricus Institoris published their infamous *Malleus Maleficarum*, better known as *The Hammer of the Witches*. Their avowed purpose was to expose what, the authors claimed, were vast networks of witches and sorcerers dedicated to serving Satan.[3] It was a time of witch-hunter generals, witch trials, witch burnings and witch hangings. An age in which mass hysteria seized people's minds and fear of damnation destroyed their reason. A period when every type of impulse was explained by satanic influences.[4] *Principia Theologica*, an anonymous 16th-century English catechism written for the growing literate middle classes, warns readers that the devil works by implanting 'base, common thoughts of an impulsive and irresistible nature' in their minds. In Germany, Martin Luther exhorted the faithful to 'avoid the impulse to harlotry', noting that 'drunkenness begins as an impulse' and warning congregations against entertaining impulses, lest they 'conquer the will'. For Luther the devil was a very real and ever-present companion. A sufferer from chronic bowel pains, the theologian firmly believed that Satan had taken up residence in his lower intestine!

From around the start of the 18th century, however, Lucifer's hold over the public imagination gradually declined and the witch trials gradually faded into history. 'Satanic influence was always a convenient answer,' comment William McCown and Philip DeSimone, 'but one that gradually lost some appeal.'[5] So long as impulsivity had been construed exclusively in spiritual terms, no progress could be made in understanding its basic causes. Indeed, the word 'impulse' does not appear on any documents until the middle of the 16th century, when it was used to describe 'bad' thoughts that arose for natural rather than devilish reasons. By the 17th century, some writers and thinkers were using the word to describe dysfunctional behaviour. Others, such as the 18th-century French writer and philosopher Jean-Jacques Rousseau, regarded it as one of the prime 'virtues of natural man': an individual who acted according to instinct and impulse rather than reason.

It was another Frenchman, the philosopher and physician Theopholi Bonet[6], who in 1684 made the first serious attempt to describe impulsivity in scientific terms. He distinguished between impulsive thoughts, obsessive thoughts, impulsive character, and the erratic and unstable moods found in manic-depressive illness. While asserting that impulses arose from mankind's 'base' nature Bonet was careful to draw a distinction between individuals capable of controlling their impulsive thoughts and other 'wretches' who 'by weakness of mind dwell on such (thoughts) without interruption'. These 'wretches', among whom Bonet numbered habitual criminals, drunkards and 'all manners of men who practice moral rights and appear to profit not from instruction', were, he contended, doomed to perpetual moral quandaries since they 'know what to do, yet fail to do so'.

The third notable Frenchman to involve himself in this debate is now recognised as one of the greatest medical reformers of his age. Dr Philippe Pinel[7] was a French physician who, during the late 18th century, laid the foundations for a more humane approach to the treatment and custody of psychiatric patients. As head of Paris's sprawling Salpêtrière mental hospital – more of a small town than an asylum with some 7,000 patients – he caused a sensation by ordering the wardens to unchain female patients. He also insisted they be treated with kindness and tolerance rather than confinement and brutality. Pinel – in common with most 'moral commentators' and physicians of his day – struggled to resolve the conflict between notions of an individual's personal responsibility for his or her actions, a function of their supposed 'free will', and the inescapable fact, which he observed daily, that many patients were incapable of 'understanding' the consequences of their actions. Moving around the crowded, chaotic wards he witnessed the distressing spectacle of patients, whose reason seemed unaffected by their madness, engaged in impulsive acts of self-harm. Even more curious, in his eyes, was the fact that those suffering from what he called *la folie raisonnante* appeared completely unable to learn from the often painful consequences of these behaviours. Such patients were, he concluded, suffering from *manie sans délire*, that is 'insanity without confusion of mind'.

By the end of the 19th century a majority of medical practitioners, including most psychiatrists, had come round to his point of view. The consensus was that those suffering from 'moral insanity' were afflicted by overpowering 'impulses' that compelled them to act in ways which they themselves would view as morally evil were they capable of doing so. In 1900, an American named Walter Dill Scott[8], who had studied under Wilhelm Wundt – founding father of experimental psychology – wrote a lengthy dissertation on the psychology of impulses. In this he sought to develop a scientific definition of the concept that avoided any of the theological and moral implications surrounding it. He proposed that the newly established discipline of psychology could make a useful contribution to separating scientific fact from religious moralising.

Two years later, in one of the first psychiatric textbooks ever published, the German psychiatrist Eduard Hirt[9] resurrected the ancient Hippocratic notion of the four humours – yellow bile, black bile, blood and phlegm. He suggested that all psychiatric conditions could be explained by reference to these bodily fluids. The personalities of patients with an overly sanguine temperament were, he considered, characterised by superficial excitability, enthusiasm and unreliability. Their principal problem, he believed, was a 'lack of impulse control'. Those with excessive yellow bile displayed a 'choleric temperament' and were angry, vociferous individuals, equally incapable of controlling their impulses. A personality formed from a combination of choleric and sanguine humours displayed excessive impulsivity, a cavalier disregard for other people and bouts of explosive anger so intense on occasions as to have them regarded as 'morally insane'. From being a spiritual manifestation of the devil's power over man, the impulse had been transformed into a subject for psychological investigation and explanation.

My purpose in writing this book is to explain what psychology and neuroscience have discovered about why we do things on the spur of the moment, usually without knowing why! I will be describing the latest research findings and showing how our baser impulses are often deliberately manipulated for commercial and political purposes. I will explore the two types of thinking that go on simultaneously in the brain, the one slow and reflective and the other fast but prone to error. The development of the brain during the first two decades of life will be examined in order to understand why teenagers behave in a more impulsive and reckless way than adults. The important role played in decision-making by heuristics, those 'rule of thumb' mental strategies that are responsible for many impulsive misjudgements, are described together with the role played by personality in risk taking. The role of the senses in triggering impulses and the nature and limitations of self-control will be investigated.

In the second part of the book, I consider in detail four key aspects of life where impulses play a major role: these are falling in love, overeating, buying things on an impulse and the impulse that leads to the destruction of oneself or others. By understanding the strengths and weaknesses of impulses, by learning to identify those occasions when we should depend on them and the times when it would be better not to do so, we may enjoy a richer and more rewarding life.

CHAPTER 1

# The Impulse That Saved My Life

'Desires and impulses are as much a part of a perfect
human being, as beliefs and restraints: and strong
impulses are only perilous when not properly balanced.'
John Stuart Mill, *On Liberty*

On 4 December 1971, while working as a journalist in Belfast, I
went to the cinema on an impulse. That impulse saved my life.

I had first visited Northern Ireland in September 1969, a few
weeks after an attempted march by the Protestant Apprentice Boys
through the Catholic Bogside area of Derry had led to three days
of rioting. Three days later, on 14 August, with civil unrest and
sectarian violence on the increase, the British government under
Harold Wilson sent in troops for what they claimed would be a
'limited operation'. It was the start of 20 years of conflict, bombings, murder and destruction that spread to the UK mainland and
came to be known, euphemistically, as 'the Troubles'. Initially
welcomed by the Catholic community as a safeguard against
Protestant violence, the soldiers soon came to be viewed as a hated
army of occupation. By the early seventies, and especially after
internment[1] was introduced in 1975, 'the Troubles' had developed
into a bitter civil war, one fought as much between the Provisional
IRA – the 'Provos' – and British troops as between the two religious
communities. The mood of the Catholic community was summed
up in a best-selling record of the time, 'The Men Behind the Wire',
written and composed by Paddy McGuigan of the Barleycorn folk
group.[2] While Protestant groups, such as the Ulster Volunteer Force

(UVF), Tara, the Shankill Defence Association, the Shankill Butchers and the Ulster Protestant Volunteers, occasionally attacked British soldiers, their violence was mainly directed against Catholics.

By the time I returned to Belfast for my sixth visit, in late November 1971, killings and bombings had become almost a daily occurrence. In the first two weeks of December alone, 70 bombs exploded, 30 people were killed and scores more were injured. During earlier visits I'd come close to a severe beating – or worse – on a number of occasions, especially when taking photographs around the Falls Road.[3] On this visit I came within minutes of falling victim to a terrorist's bomb myself.

## An appointment with the McGurks

Among the many friends I had made, on both sides of the sectarian divide, was Dr Jim Ryan. A Catholic GP, Jim had devoted his life to caring for the poor in slums around the Catholic Falls and the Protestant Shankill Roads, irrespective of their religion. At lunchtime on that bright but chilly December day, I met up for a drink with Jim in a bar called Kelly's Cellars, the city's oldest licensed premises.[4] I told him that my latest assignment was to write about the effects of the violence on Belfast's young people and asked whether he knew of any families with teenagers whom I could interview. Jim immediately suggested I talked with the McGurks.

Patrick and Philomena McGurk ran a public house called Tramore, known to locals as McGurk's, which stood on the corner of North Queen Street and Great Georges Street. Although this was in a staunchly Catholic and Irish nationalist area of the city, Patrick and Philomena were well known for their lack of religious bigotry. They also had a bright, attractive and eloquent 14-year-old daughter named Maria. I gratefully accepted Jim's suggestion and he arranged for me to meet mother and daughter at around 8.30 that evening.

I left my hotel shortly after seven to have a meal before going on to my meeting. It was a chilly night and the darkened streets of the city looked especially depressing and foreboding. On the spur of the moment I decided to go to the cinema. The film ended just before nine. I found a taxi easily enough and told the driver to take me to Great Georges Street. We never arrived at our destination.

Approaching the junction with North Queen Street it became clear a major terrorist incident had just taken place. The chilly night air was filled with dense smoke. I could smell the acrid fumes of explosive and burning timbers. The street was filled with fire engines, ambulances, police cars and army trucks. Eerily lit in the harsh white glare of emergency lamps and the occasional flash of a press photographer's camera, soldiers and civilians scrambled and crawled over the avalanche of rubble, digging and burrowing desperately amongst charred timbers and fragmented brickwork. Of McGurk's scarcely anything recognisable remained. A solitary, soot-blackened wall. A lone metal arch standing out amidst shattered timbers and twisted steel. The explosion had flattened the old building like a giant's foot crushing a child's toy.

At almost the exact moment I left the cinema, a bomb disguised as a brown-paper parcel had exploded in the entrance to the crowded bar. Seventeen people had been killed. Among them Philomena and her daughter. Maria had been doing her homework in the pub's first-floor living room when the bomb exploded. She was killed instantly. If I had been with them, I too would almost certainly have died.[5]

I will discuss the possible reasons for that life-saving impulse in a moment. But let's first look at three other examples of people who acted on an impulse and survived almost certain death as a result.

## The man who was never late

Fifty-four-year-old Fred Eichler, Chief Financial Officer with the New York brokerage firm Axcelera, prided himself on always being punctual. That bright, sunny autumn morning was no exception. A few minutes before 8.15 he stepped into his building's express elevator and rode it to his office on the 83rd floor of the World Trade Center's North Tower. The date was 11 September 2001 and the lives of more than 2,800 men and women in the offices around him were about to be brutally ended. At 8.40 he left his desk to go to the lavatory. On his way he met a group of colleagues and made a sudden decision to stop in a nearby conference room for a chat. While talking, their eyes were suddenly transfixed by the sight of a Boeing 767 passenger jet heading straight towards them through the cloudless blue sky. One of them said in alarmed amazement: 'Gee, that plane is flying awful low!'

'It must be a plane from Kennedy that's got into trouble,' Fred replied.

Afterwards he recalled: 'It was all in slow motion. I am told that the plane was flying towards us at 600mph, yet it seemed like an eternity getting to us. I suppose it was 15 seconds. None of us really expected it to hit the building. But it just kept coming and coming. Most of the time it was . . . right in line with the window we were staring out of. Then it was almost on us. I could make out the seams on the wings and all the American Airline markings. I looked right into the cockpit but I couldn't really make out the figures. They were tiny windows and the sun was shining on them.

Maybe I eyeballed Mohammed Atta, the hijack pilot, but I can't be sure.'

When it was within 200 yards of the building, the aircraft reared up suddenly and banked abruptly to the right. Fred now found himself staring in astonished disbelief at the Boeing's gleaming silver belly. A moment later, at 8.46 and 26 seconds, the tip of the wing struck the offices some 70 feet above them. It ploughed through floors 94 to 98, smashing steel columns, shredding metal filing cabinets and crushing computer-laden desks. Almost instantly the aircraft's 10,000 gallons of fuel ignited into a vast fireball, incinerating everything and everyone in its path. So massive was the impact that the aircraft's landing gear was flung through the south wall of the tower to come crashing down on a street five blocks away. The shock wave from the initial explosion flung Fred and his companions to the ground. Flames and dense black, acrid smoke billowed through the corridors outside the room. Had he still been standing there, Fred would have been killed instantly. But it was not until much later that he appreciated how narrowly he had escaped death. Despite the fact that more than a thousand are thought to have survived the initial impact, not a single person working only seven floors above him got out of the building alive. In the inferno that swept through the North Tower they were either burned alive, suffocated by smoke, killed when the building collapsed or driven by the unbearable heat to leap to their deaths. 'No one could comprehend it,' says Fred. 'I still can't.'

Corridors filled with dense smoke and flames, shattered stair-wells filled with water cascading down from broken sprinkler pipes made escape from the doomed building far from straightforward. In the hallway immediately adjoining the conference room where Fred and the others were taking refuge, the fires had been beaten back by jets of water from the sprinklers and the lights still burned brightly. At 9.30 they suddenly saw a flashlight held by a fireman accompanied by a building worker. Guided by the fireman they made their way down the stairs, stumbling through vast puddles of water and passed by firefighters making their way up towards the blazing floors above them – and to their deaths. As they reached the ground floor a lift shaft shattered into fragments and came crashing down. Ahead of him Fred saw a broken window and scrambled through it to the street. Four minutes later, at 10.28 a.m., the North Tower collapsed. 'I can't get away from it,' Fred says sadly. 'On our floor there were 15 people killed, ten seriously burned – one of them in the men's room. If I had gone in there, where I had been heading at the time, I might not be here now.'[6]

Both Fred Eichler and I undoubtedly owe our lives to an impulse. But in neither case was this the result of any sense of impending danger. The next two stories, however, are different. In each an intense if inexplicable fear motivated boy and man to act in the way they did.

## The boy in the tunnel

At one stage of my career, while lecturing in clinical psychology and psychopathology at the University of Sussex, I was running a registered charity – Action on Phobias – which helped people suffering from a range of anxiety and phobic difficulties. I also saw patients privately and the two accounts below, with names and certain minor details changed to protect confidentiality, illustrate how a subconscious sense of danger can trigger life-protecting impulses.

Eleven-year-old Tony and his 14-year-old brother Michael lived on a West Country farm managed by their father. The farm had a very large Dutch barn filled, during the summer, with hundreds of straw bales. The two lads had constructed a secret camp by hollowing out a space between the bales in the centre of the barn. This could only be reached through a long, narrow passage between the bales. It was just high enough for a small boy to crawl through on his hands and knees. One night they decided to spend the night in their secret hidey-hole. As darkness fell, Michael wriggled his way in first, pushing their sleeping bags before him. Tony was about to follow when Michael reappeared saying he had left the food for their midnight feast behind in the kitchen and would run to fetch it. Tony decided to go ahead without him. Thirty-four years later he could still remember every detail of scrabbling on all fours along the darkness of the tunnel.[7] The straw pricking his bare knees, scratching at his arms and brushing his head as he crawled through the musty, straw-scented pitch blackness to their hideaway deep in the barn. He had gone about ten metres down the tunnel when he froze abruptly, suddenly incapable of moving forward. Rooted to the spot by an inexplicable sense of danger. The onrush of fear surprised Tony. He had never previously felt fearful while in the tunnel and was not in the slightest way claustrophobic. On an impulse, and propelled by this unexpected sense of dread, he began a hasty retreat. Since there was no way he could turn round in the narrow passage he had to crawl out backwards. When he was about three metres from the fresh air he saw an orange glow at the far end of the tunnel. A glow that raced towards him at lightning speed preceded by a wall of heat more intense than he had ever experienced. Tony could vividly remembered seeing, in stark detail, the straw forming the roof and walls of the tunnel, all brilliantly illuminated as by an arc lamp. He later learned that his brother had taken a candle into the secret hideout and left it burning while he went back to fetch the food. It must have ignited the straw and started the blaze that destroyed the barn. As Tony scrambled, now in a state of total panic, into the moist night air, a geyser of flame and smoke erupted from the tunnel, singeing his hair and burning his face. Within moments the barn was a blazing inferno. Had he not retreated when he did there can be little doubt that he would have been incinerated by the blaze.

## The man who didn't change trains

Peter came to see me suffering from what, today, we would call post-traumatic stress disorder. A mixture of severe anxiety mixed with guilt, it was the result of a remarkable escape from death he had experienced a year before. On the evening of 18 November 1987, 32-year-old Peter had left his office in central London slightly later than usual and hurried to catch the Piccadilly Line train for the first stage of his journey home. He would travel to King's Cross station then change to the Circle Line. It was a commute he had done hundreds of times before. The train pulled into King's Cross shortly before 7.30 and Peter hurried onto the escalator. This would take him up to the ticket hall from which he could transfer to the Circle Line.

The main-line station was, as usual, extremely crowded but everything appeared perfectly normal. As the escalator began to take him towards the ticket hall, however, Peter was suddenly overcome by an overwhelming impulse to go back the way he had come. So intense was this feeling that, ignoring the angry protests of other passengers, he turned around and forced his way back down the escalator to the platform he had left only moments earlier. A train was just pulling out. He jumped on it and collapsed panting into a seat, completely unable to explain his irrational behaviour.

A few seconds after he left the station it was engulfed by fire. Within minutes the whole place was a raging inferno. The ticket hall and the escalator on which he had been travelling were destroyed and 31 people were killed. Many who died were those ahead of and behind him on the escalator from which he had so precipitously and inexplicably fled.

How can we explain such fortuitous impulses? In my own case, I had no particular premonition of danger above the nagging sense of foreboding that was my constant companion while in Belfast in those days. Perhaps it was the contrast between the dark and gloomy December streets and the brightly lit cinema foyer, cheerfully decked out for Christmas, that caught my eye. Maybe it was the prospect of being able to lose myself in a film – of which I can no longer recall even the title, let alone any details of the plot, such was the impression it made – that was so inviting. Certainly compared to the alternative of a solitary supper in a drab restaurant. In his powerful and moving account of escaping from the North Tower, Fred Eichler makes it perfectly clear there was nothing untoward in his mind when he decided to delay a visit to the restroom in order to chat with colleagues. As for Tony and Peter's life-saving impulses, although they were unable to explain their behaviour, my hunch is that they were responding to the smell of danger, albeit below their level of conscious awareness. They responded instinctively, using what I describe in the next chapter as their 'zombie brain', to subtle and virtually imperceptible indications of danger. The fire started by the older boy's candle, for example, would have been burning for a while before it produced that gush of flame that chased Tony down the tunnel. Similarly, reports show that the fire at King's Cross station initially started below the escalator. Probably the consequence of a carelessly dropped match or lighted cigarette, it had been burning for some while prior to engulfing the whole place in flames. The fire brigade, who were already on the scene when Peter's train pulled into the station, believed it posed little or no danger to the public. Although Peter, possibly because he was so focused on getting home, had no recollection of seeing anything untoward prior to his sudden decision to flee, subconsciously he may have smelled, seen or heard something that triggered an alarm bell deep in his brain.

Now of course, had McGurk's not been bombed that night, if a fallen candle had not turned the Dutch barn into an inferno and had the tragedy at King's Cross never occurred, then Tony, Peter and I would probably have quickly forgotten our actions. Tony might have been momentarily ashamed of his loss of nerve and probably teased about it by his older brother. Peter might well have occasionally recounted his precipitous flight down the upward-moving escalator with an embarrassed and self-deprecating chuckle. It is most unlikely, however, that any one of us would have long remembered the incident in such vivid – if possibly inaccurate – detail. Which brings me to an important point about impulses. They only ever become significant in the light of subsequent events. Take, for example, the happier impulse that led grandmother Maire McKibbin, from Kilkeel, Co Down, to buy five lottery tickets rather than her usual one. She won over a million euros[8] with the fifth ticket, making her spur-of-the-moment decision both memorable and newsworthy. Had she had failed to win anything Marie would very likely have forgotten all about it. As William James commented in his 19th-century book *The Principles of Psychology*, 'In such cases the *strength of the impulsive idea is preternaturally exalted*, and what would be for most people the passing suggestion of a possibility becomes a gnawing, craving urgency to act.'[9]

In this book I argue that virtually everything we say and do between waking and sleeping can be considered impulsive in that the vast majority of those actions occur *mindlessly* rather than *mindfully*. They result from mental processes operating below conscious awareness. If correct, this means that when it comes to giving in to our impulses we are behaving not as rational human beings but as zombies.

# CHAPTER 2

# Impulses and Your Zombie Brain

'There is a zombie within you that is capable of
processing all the information your conscious self can
process consciously, with one crucial difference . . . your
zombie is unconscious.' Anthony Atkinson, Michael
Thomas and Axel Cleeremans, *Trends in Cognitive
Sciences*[1]

Visitors to my laboratory, on the south coast of England, often
feel disgusted. Or they may become angry, anxious, amused,
astonished, frustrated, bewildered, stressed, startled, shocked or
sexually excited. It's all done with their full consent and in the
name of science. Our studies are designed to investigate the role
of emotions in triggering impulsive behaviours. A typical research
project involves 'wiring up' our volunteers to record what's going
on in their brain and body during the experience. A network of
electrodes are attached to their scalp and connected to an EEG
(electroencephalograph) that records electrical activity in their
brain. This will indicate the extent to which they are relaxed, by
the predominance of relatively slow brain frequencies – known
as alpha waves – or cognitively aroused and generating the faster
beta waves. Other sensors record changes in their heart rate,
respiration and skin conductance. Increases in the latter indicate
rising levels of stress. A video camera, focused on their faces,
will produce a slow-motion record of their expressions throughout
the research, while infrared detectors will track their eye movements on the screen.

When the study starts we see brain activity shifting abruptly, the slower alpha waves being replaced by beta waves as brain activity increases. There is an accompanying rise in heart rate, skin conductance and breathing, the latter typically becoming faster and shallower. To get a flavour for the type of images our participants see, take a look at the two stills in figure 1 from a video depicting self-mutilation.

*Figure 1: Two views of self-mutilation*

How do the images make you feel? Intrigued or upset by the sight of someone slicing into their arm with a razor-sharp butcher's knife? Perhaps you glanced away or turned the page. Maybe you examined the pictures to discover whether or not they really do depict man in the act of vicious self-injury. If so, you were right to be suspicious: the arm is my own, the knife a trick one and the 'blood' came out of a bottle!

Although the intensity of the physical responses in experiments like these varies from one person to the next, the physical responses occur almost instantly. When shown sickening images, of which the 'hand chop' is fairly low down on our list of horrors, they typically take a sharp intake of breath, sometimes clamping a hand over their mouth as their eyes widen. These impulsive responses typically occur in situations that combine high levels of emotion, whether positive or negative, with uncertainty about what is being seen or heard, tasted or touched. They are an example of what I have termed System I (for impulsive) thinking in action. It operates rapidly and automatically without our ever being aware of what is going on behind the scenes. It contrasts with System R (for reflection), which is the slower, more methodical, conscious reasoning we use when trying to solve a challenging but unfamiliar problem or reach a decision outside the normal run of decision-making.

## Most of our thinking happens backstage

In 1896, two American psychologists, Leon Solomons and Gertrude Stein, were studying a phenomenon known then as 'double personalities' and later as 'split personalities'. They suggested there was a similarity between the 'mindless' actions of the second personality and those of 'normal' men and women who engage in a wide range of behaviours without consciously attending to them. Acting as their own guinea pigs, the two performed a great number of experiments, in which they showed it was possible to learn how to both read and write simultaneously. After much practice they were able to write coherent sentences at speed while simultaneously reading and absorbing details of a complicated story. Later they even mastered the seemingly impossible skill of taking virtually error-free dictation as they carried on reading. In one study they showed that it was possible to read aloud from a book while also paying close attention to a story being read to them.

By developing these strange and unusual skills Solomons and Stein were able to demonstrate that a great many actions we regard as requiring intelligence, such as reading and writing, can be carried out quite automatically. 'We have shown a general tendency, on the part of normal people, to act, without any express desire or conscious volition, in a manner in general accord with the previous habits of the person,' they commented presciently.[2]

Today neuroscience has confirmed what, for many, seems both remarkable and startling. The fact that the majority of our thinking goes on without our ever being aware of the fact! As cognitive psychologists George Lakoff and Rafael Nunez point out: 'Most of our thought is unconscious – that is, fundamentally inaccessible to our direct, conscious introspection. Most everyday thinking occurs too fast and at too low a level in the mind to be accessible. Most cognitions happen backstage.'[3] When Venus Williams slams a ball across the net she does so by merging conscious attention and unconscious perception into a seamless movement, knowing exactly where to intercept the ball and the precise angle at which to hold her tennis racket. But ask Venus Williams to describe the calculations her brain performs in the split second between seeing the tennis ball hurtling towards her and returning it to her opponent and she would be lost for words.

One way of disrupting such 'automatic' behaviour is by asking the person to think about what they are doing. Try this the next time you find yourself being hammered into the ground by your opponent at tennis. Just say, in apparent admiration: 'That was a brilliant serve, how on earth did you manage to do that?' The chances are that, from then on, your opponent will start thinking about how they are serving and so end up producing nothing but double faults!

System I thinking, which we share with other animals, operates outside conscious control. It works by developing categories and automatically placing events, people, actions and situations into them. It comes into its own when we need to respond to fast-moving events. Rapid, emotional and generalised, it enables us to react, instantly and automatically, in ways we are only rarely able to explain. It is also very easily fooled.

Reflective, or System R, thinking is under our conscious control. It is rational, logical and sceptical, constantly asking questions and seeking answers. It analyses, plans, computes, predicts and strives – often with little success – to regulate the thoughts, words and actions produced by its impulsive companion. It enables us to 'think about thinking', to indulge in abstract, hypothetical reasoning, to plan, to predict and to foresee, to construct mental models and create imaginary futures. It endows us with a unique potential for a higher level of rationality when problem-solving and decision-making. Reflective thinking is conscious thinking. It is slow, analytic, sequential, controlled and linked to language. It has a low processing capacity, makes heavy demands on memory and requires a high level of effort.[4]

System R coexists with the evolutionarily far older System I mode of thought. Sometimes the two systems collaborate, sometimes they are in conflict and on other occasions they merge seamlessly into one. What starts as an impulse-driven idea may segue into profound reflective thought. What seems at first sight to be the outcome of reflective thinking turns out, on closer inspection, to be based on an impulse. To get a flavour of these two systems, consider the following examples of System I and System R in action.

## System I – fast and intuitive

Bob Nardelli's boyhood dream of becoming a professional footballer was dashed when he was still a teenager. Not by any lack of determination on his part but by his stature! At 5 foot 10 and 13 stone 13 pounds he was the smallest player on the line at Western Illinois University. 'The rest of the world got bigger, and I didn't grow any more,' he recalls.

Setting aside his youthful ambition, Nardelli worked hard to gain a business degree before taking a job on the factory floor at General Electric, the company where his father had spent a lifetime. Slowly but surely he climbed the corporate ladder, working the longest hours, taking on the toughest challenges and consistently delivering his projects on time and on budget. He was, in the words of his company's CEO, the legendary Jack Welch: 'The best operating executive I've ever seen.' But when, in November 2000, Jack finally stepped down, his final decision as CEO was to promote not Bob Nardelli but his rival Jeff Immelt to the top job. Shocked and disbelieving, Nardelli demanded to know why he had been passed over. 'Tell me what I could have done better. Tell me that the numbers weren't there, the innovation, the talent development, the relationships with Wall Street. Give me a reason.' All Welch could say was: 'It was my call. I had to go with my gut.'[5]

## System R – slow and methodical

In the autumn of 1940, as Britain braced itself for the German invasion, the intelligence services intercepted some unexpected and bizarre signals. The intercepts were sent to Britain's Code and Cipher School at Bletchley Park, some 60 miles north of London. There, cryptanalyst Brigadier John Tiltman recognised the signals as coming from an advanced German teleprinter, the Lorenz machine. Named after its Berlin manufacturer, the Lorenz machine was capable of both enciphering and transmitting messages or receiving and deciphering them.[6] The breakthrough the Allies were hoping for occurred on 30 August 1941, when a German army operator based in Athens was given the unenviable task of sending a 4,000-character-long message to his colleague in Vienna. Typing such a lengthy message was a tedious business and the operator must have been disheartened to be informed by his opposite number in Vienna: 'We didn't get that! Send again.' Perhaps due to frustration, the operator made a careless blunder. Instead of resetting the machine, as regulations demanded, he saved time by using the initial settings. Had this been his only mistake the Bletchley team would merely have intercepted two identical versions of the message. However, as he laboriously typed in the message for a second time the German soldier introduced some minor changes. As a result the previously secure Lorenz machine was turned into a floodgate through which, for the next four years, German High Command's deepest secrets poured forth.

Not that capitalising on the German operator's carelessness was a simple task. It took eight weeks of intense mental effort by a Bill Tutte, a young chemistry graduate only recently recruited from Cambridge, to crack the cipher. In the days before computers he achieved this intellectual tour de force using nothing more sophisticated than a pen and some paper. He painstakingly compared the two sets of data streams until a pattern emerged. When it did, not only were the contents of individual messages revealed but so too was the logical structure of the supposedly 'unbreakable' Lorenz system.[7]

In the two examples above, Jack Welch relied on intuitive thinking, on what he described as his 'gut' feelings, while Bill Tutte employed rational, logical thinking of a high order to extract meaning from the apparently incomprehensible data. Tutte was, at all times, fully aware of his chain of reasoning. He knew what was going on in his mind as he struggled to break the cipher. Welch had no idea why he came down in favour of Jeff Immelt rather than Bob Nardelli. Most of the time our thought processes resemble those of Jack Welch far more than they do Bill Tutte's.

## Reflective versus impulsive thinking

The ease and frequency with which System I overrides System R was illustrated for me by the tragic case of a bright young student in my tutorial group while I was lecturing in clinical psychology and psychopathology at the University of Sussex. Two years earlier, he had gone to Greece on holiday with a group of friends. One afternoon, as they were strolling on a cliff top high above the sparkling Aegean, he impulsively dived into the water below. He struck a submerged rock, barely visible beneath the shimmering surface, and shattered his spine. A split-second impulse left him paralysed from the waist down for the rest of his life. Stupid and thoughtless were two of the words used to describe that impulsive leap, not least by the young man himself. Yet, had he splashed safely into the ocean, his daring would have won him admiring glances from the girls and envious looks from the boys.

'In the routine of daily life we do not notice what we are doing unless there is a problem,' comments Ellen Langer, Professor of Psychology at Harvard University. 'The consequences . . . range from the trivial to the catastrophic.'[8] Professor Langer has spent a lifetime researching what she calls 'mindfulness' and setting up simple but compelling studies to illustrate how often this is overwhelmed by impulse-driven mindlessness. On one occasion she sent a memo around university offices stating bluntly: 'This memo is to be returned to Room 247.' Half the memos were identical to those typically sent between departments. The other half were slightly different. Clearly, anyone employing System R thinking while reading the instruction would have immediately asked: 'If whoever sent this memo wants it back why was it sent to me in the first place?' But this is not what happened. When the memo looked official, nine out of ten recipients did as ordered and promptly returned it. Even when it looked slightly unusual it was returned on six out of ten occasions.[9]

In another experiment Langer and her colleagues decided to test whether they could persuade people using the library photocopier to let them jump the queue.[10] The setting for this unusual piece of research was the Graduate Centre at the City University of New York. The person conducting the experiment sat at a table giving them an unrestricted view of the photocopier. As soon as anyone went to use the machine, they were approached by the researcher who asked to use the copier ahead of them. Their request took one of three forms. In the first version the interloper simply said: 'Excuse me, may I use the Xerox machine?' This was the 'request only' approach. In the second version the person was asked: 'Excuse me, may I use the Xerox machine, because I have to make copies?' Here the request was accomplished by an apparent explanation which was, in fact, meaningless! After all, who uses a photocopier except to make copies? The third approach combined a request with a believable explanation: 'Excuse me, may I use the Xerox machine, because I'm in a rush?' As well as varying the nature of their request the researchers also asked either a small favour 'May I make five copies?' or a bigger one 'May I make 20 copies?'

'We gave reasons that were either sound or senseless,' Langer reports. 'An identical response to both sound and senseless requests would show that our subjects were not thinking about what was being said.'[11] The first of the two versions with explanations, '. . . because I have to make copies', would strike anyone using System R thinking as absurd. Langer, however, hypothesised it would seem perfectly sensible and reasonable to those using System I thinking. She was right! Provided the favour was a small one (five copies) 60 per cent of those approached with the 'request only' version willingly yielded their place. When the request was followed by either a senseless ('because I have to make copies') or a sensible ('because I'm in a rush) explanation compliance rose to 93 per cent for version two and 94 per cent for version three. When the favour was a big one (20 copies) only the explanation based on sensible information – 'because I'm in a rush' – made any difference to levels of compliance, with 42 per cent agreeing to the request. This compared with 24 per cent approached either without any explanation or with the senseless version. While significantly smaller, the portion who impulsively agreed to pass up their turn is still remarkable.

What is going on here? 'We become mindless . . . by forming a mindset when we first encounter something and then clinging to it when we re-encounter that same thing,' Ellen Langer explains. 'When we accept an impression or a piece of information at face value . . . that impression settles unobtrusively into our minds . . . most of us don't reconsider what we mindlessly accepted earlier . . . the mindless individual is committed to one predetermined use of the information, and other possible uses or applications are not explored.'[12]

## System I thinking and your zombie brain

Look hard at the illustrations in figure 2 and something odd will happen.

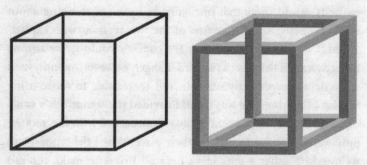

*Figure 2: Necker Cubes*

As you study the cubes – which are popularly known as Necker Cubes after the Swiss crystallographer Louis Albert Necker, who discovered the phenomenon – they will abruptly change shape.[13] You may find, for instance, that at one moment the lower right edge appears to be located at the front of the cube and a moment later at the rear. If you find that this is not currently happening try blinking. You will discover that you have no conscious control over when the switch occurs, nor any ability to prevent it happening.

We also use this automatic system when engaged in routine mental tasks. If, for example, I asked you to add 2 and 2 you would automatically answer 4. If I asked you to divide 228 by 19 in your head, however, you would probably try to visualise the numbers mentally (System R) to come up with the answer 12. Both these are instances of your non-conscious 'zombie brain' at work. Our zombie brain seeks to make sense of the world around us, as well as of our own internal milieu, by creating categories, 'mindsets', which are like small boxes into which incoming information can be sorted. By compartmentalising it we are able to make rapid, if crude, decisions and reach swift, if often wildly inaccurate, conclusions. When the information the zombie brain receives is ambiguous, as in the illusion above, it offers two equally probable interpretations to the conscious mind, shrugs its metaphorical shoulders and says, in effect: 'Your guess is as good as mine!' Zombie thinking can easily result in a problem termed 'functional fixedness': a narrowing of possible options that limits our creativity and problem-solving capacity.

In a classic early study of this phenomenon, the early twentieth-century German psychologist Karl Duncker gave volunteers three objects: a candle, a box of drawing pins and a book of matches, as shown in figure 3.

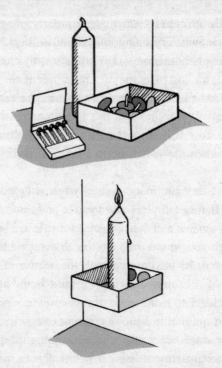

*Figure 3: Candle, box of drawing pins and book of matches*

Their task was to attach the candle to the wall in such a way that it did not drip onto the table below. He found a majority of his participants either attempted to attach the candle to the wall with the drawing pins or stick it there by melting some of the wax.[14] Only a small minority ever thought of using the inner compartment of the box as a candleholder and attaching this to the wall with a couple of the drawing pins. When, in subsequent experiments, participants were provided with an empty box they were twice as likely to solve the problem. Duncker explained this failure by suggesting partici-pants had become so 'fixated' on the usual purpose of the box, that is as a container for drawing pins, they were unable to view it in a way that enabled them to solve the problem. They created a System I 'mindset' that inhibited creativity. While mindsets speed up thinking by encouraging us to use impulse-driven responses – and can occasionally prove a life-saving mode of thought, at the other extreme they can also lead to actions ranging from the foolish to the fatal.

## Functional and dysfunctional impulses

In 1990 Scott J. Dickman, then at the University of Texas, Austin, distinguished between two different types of impulsivity, which he called *dysfunctional* and *functional*.[15] Dysfunctional impulses are those generally considered to be thoughtless, reckless or self-destructive. Condemned by others, their outcomes are always negative and occasionally fatal. Functional impulses, by contrast, are usually seen as bold, daring and courageous. Praised by others, their outcomes are usually positive and rewarding. Scott Dickman found that these two types of impulsiveness bore 'different relations both to other personality traits and to the manner in which certain basic cognitive processes are executed'.[16] While the outcomes of functional and dysfunctional impulses can be very different, those differences frequently owe far more to good luck than good judgement.

## Functional impulsivity – the bus that flew

On the morning of 30 December 1952, a Number 78 double-decker bus, with 20 passengers and a conductor on board, 'flew' over the River Thames. Not that the driver, 46-year-old Albert Gunter, ever intended to do so when he set off on his six-mile journey through London. It was a route he knew all too well, having driven it,

without incident, hundreds of times before. That day his journey
would turn out to be very different. Part of his trip involved
crossing the river via Tower Bridge. From time to time the bridge's
two massive 1,000-ton spans were raised to allow a ship to pass
beneath them. Before this happened, clear procedures had to be
followed. A gateman rang a warning bell and two barriers came
down to stop traffic. Only once this was done, did the duty
watchman order the bascules to be raised. That morning, with a
relief watchman on duty, the safety procedures spectacularly failed.
As Albert Gunter's bus trundled over the south bascule, the
100-foot span began, without warning, to tilt sharply upwards.
Tower Bridge had started to open! 'It seemed as though the road
in front of me was falling away,' Albert Gunter later recalled.
'Everything happened terribly quickly, I realised that the part we
were on was rising. It was horrifying!'

The bus driver was left with only two options and no time to
think about either. He could slam on the brakes and hope the bus
would stop before it plunged 60 feet into the murky grey river
below. Or he could accelerate hard and find out whether his 12-ton
bus could fly. With no time to engage in System R thinking, Albert
Gunter's 'zombie brain' kicked into action and he thrust down as
hard he could on the accelerator. Gathering speed, the bus roared
up the steeply inclined span. The next instant, driver, conductor,
screaming passengers and bus were all hurtling through space.
Albert glimpsed the swirling grey waters of the Thames flash
beneath his wheels as the red double-decker flew across the yawning
gap. Moments later it landed heavily on the opposite bascule which,
rising more slowly than the first, was still some six feet lower. The
bus broke a spring, the conductor a leg and 12 of the 20 passengers
suffered minor injuries. But Albert Gunter's impulsive response
had almost certainly saved all their lives.

'Quick thinking', 'daring', 'skilled' and 'courageous' were some of the words used by press and public to describe his action. The bus company bosses were so delighted they awarded Albert a £10 bonus – about £230 in today's money – for his remarkable piece of driving. But it could all have ended so very differently, as the following account of an equally impulsive decision illustrates.

## Dysfunctional impulsivity – death of a mountain man

On 12 February 1995, three experienced skiers set off on a trip to the slopes of Utah's Wasatch mountain range. The group's leader, 37-year-old Steve Carruthers, was a veteran cross-country skier intimately familiar with the terrain. A couple of hours after setting out, Steve and his companions met another party of skiers and they discussed the best route through the mountains. A storm, 24 hours earlier, had deposited two feet of new snow on the slopes and now it was becoming increasingly foggy. The fast-deteriorating conditions caused some of the skiers to take a more prudent route across the lower slopes and they urged the others to do the same. On an impulse, Steve ignored their advice and, with two companions, chose instead to follow a more precarious route along the tree-lined higher slope. An hour later he was dead. His group inadvertently triggered an avalanche that sent hundreds of tons of snow roaring down the mountainside at more than 50 miles an hour. Unable to ski clear, Steve was crushed against the trunk of an aspen and buried deep beneath the snow. Unconscious when finally dug free, he never regained consciousness and died in the air ambulance on his way to hospital. The two other skiers survived to face a barrage of criticism from the media, the authorities and the tight-knit skiing community.

Why, since the pass was well known avalanche terrain and February a high-hazard season, had such a knowledgeable cross-country skier like Steve acted so impulsively? What had led a man with more than 20 years' experience on those same slopes to place himself and his party in jeopardy through so reckless an error of judgement?[17] The answer seems to be that his zombie brain, using System I thinking, had led him to act on an impulse rather than after more careful reflection. But if the story had had a happier outcome then no one would have condemned his decision or criticised him for lack of care. The same can be said of bus driver Albert Gunter's Thames leap. A successful outcome meant that his impulse was seen as daring and praiseworthy. Had it ended with the bus plunging into the Thames and killing all aboard, it would almost certainly have been condemned as reckless, stupid and perhaps suicidal.

In summary, System I thinking, our zombie brain's default mode, controls most of our everyday behaviours. Only if asked why we spoke or acted as we did do we engage System R thinking to try and come up with a sensible, or at least plausible, explanation for our behaviour. In this role it serves as a PR department for the 'self', explaining, embellishing and justifying our actions both to ourselves and to others. This task is often complicated by the fact that we may have no rational explanation for what we said or did and can only fall back on the lame excuse that it 'seemed like a good idea at the time'. In chapter 5 we will be looking at some of the ways our surroundings can trigger impulses. These triggers often act so subtly that we are not even aware of the impact they are having or how they are influencing the way we behave.

# CHAPTER 3

# Inside the Impulsive Brain

'The human brain is like nothing else. As organs go, it is not especially prepossessing . . . rounded, corrugated flesh with a consistency somewhere between jelly and cold butter. It doesn't expand and shrink like the lungs, pump like the heart, or secrete visible material like the bladder. If you sliced off the top of someone's head and peered inside, you wouldn't see much happening at all.'

Rita Carter, *The Human Brain*

At 4.30 in the afternoon of Wednesday 13 September 1848, a bizarre accident propelled an American railway worker named Phineas Gage from 19th-century obscurity to 21st-century celebrity.[1] A photograph of 26-year-old Gage, taken a few months after the accident, shows a soberly dressed young man with a slightly lopsided gaze. His left eye is closed as a result of his accident but other scars are concealed beneath dark, neatly brushed hair. In his hands he holds the long iron rod that was driven through his skull by the explosion. This is the 'tamping rod' that he had made especially for his work on the railway and which he took with him to the grave. That winter's afternoon, Gage was in charge of a small team of workers engaged on railway construction south of Duttonsville, a village in the Vermont township of Cavendish. Their task was to blast rocks out of the path of the advancing track-layers. As foreman his job was to tamp[2] or pack explosives into holes drilled in the rock using an iron rod. It was a straightforward task Gage had undertaken

many times before. First he would place an explosive charge in
the hole and then gently pat down an inert material around the
charge to hold it in position. Next he inserted a fuse before filling
the hole with sand or clay. This was then compressed by firm
use of the long tamping rod. While tamping demands no great
skill, it is work that must be carried out attentively and carefully
to avoid striking a spark and prematurely exploding the charge.
On that fateful afternoon, this is exactly what occurred.

Twenty years later Dr John Martyn Harlow, the official railway
physician who attended Gage shortly after the accident, presented
a paper to a packed meeting of the Massachusetts Medical Society.
Entitled 'Recovery from the passage of an iron bar through the
head[3], it offers a graphic account of the accident and the injuries
that resulted. After a brief introduction from the chairman, Dr
Harlow – a slender, distinguished-looking professional man with
a neatly trimmed beard, moustache and greying hair above a wide
brow – rose to speak. He began by describing Gage's character,
explaining that he had been a responsible and thoughtful young
man regarded by his employers as an 'efficient and capable'
foreman. Having set the scene, Harlow took his audience back to
the afternoon of his accident. 'The powder and fuse had been
adjusted in the hole, and he was in the act of "tamping it in", as
it is called, previous to pouring in the sand,' he explained. 'While
doing this, his attention was attracted by his men in the pits behind
him . . . at the same instant dropping the iron upon the charge,
it struck fire upon the rock, and the explosion followed, which
projected the iron obliquely upwards, in a line of its axis, passing
completely through his head, and high into the air, falling to the
ground and several rods behind him, where it was afterwards
picked up by his men, smeared with blood and brain.' Harlow
continued in quiet, measured tones, describing how the tamping
rod had been thrust violently upwards through Gage's left

cheekbone. It had then penetrated the base of his skull, just behind the left eye socket, before finally exploding through the top of his head, smeared with blood, brain tissue and shards of bone. 'The fragments of bone are being lifted up, the brain protruding from the opening and hanging in shreds upon the hair . . . the edges of the scalp everted and the frontal bone extensively fractured, leaving an irregular oblong opening in the skull of 2″ × 3.5″. The globe of the left eye was protruding from its orbit by one half its diameter, and the left side of the face was more prominent than the right side.' Harlow glanced around at his colleagues' astonished faces. He knew the one question on everyone's lips was: 'How in God's name did Gage survive?' It was a question to which the doctor had no answer.

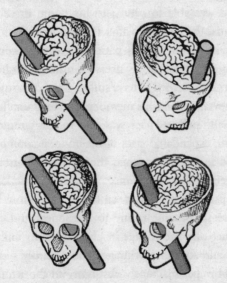

*Figure 1: The tamping rod as it passed through Gage's skull*

He described how after the accident Gage, who had remained fully conscious throughout his ordeal, was helped to stand and walked the few hundred feet to an ox cart. There he was assisted onto the seat beside the driver and sat bolt upright while being driven to his nearby lodgings in the local tavern. 'On arrival,' Harlow continued, 'he stood up by himself, walked to the back of the cart, and allowed two of his men to help him down. They helped him up the steps onto the veranda where he sat himself on a chair to await medical attention.' Despite the foreman's optimistic insistence that he would be 'back at work within a day or two' neither of the two doctors who came to see him thought the patient would live. Indeed, so gloomy was their prognosis that they sent for Thomas Winslow, the local undertaker, to measure him and prepare a coffin 'in readiness to use'.

Against all expectations and all odds Phineas Gage made a full recovery and was able to return to his home in Lebanon, New Hampshire just three months after the accident. Harlow modestly refused to take credit for his patient's recovery, concluding his presentation by insisting 'I dressed him. God healed him.' Remarkably, Gage seemed to have suffered no impairment of either speech or movement and his memory remained unaffected. While Harlow initially believed this was due to the tamping rod only destroying an 'expendable' part of his brain, it soon became clear the man had lost far more than the doctors at first realised. To Harlow it seemed that the balance between 'his intellectual faculty and animal propensities' had been taken away. From being polite and reflective he became prone to selfishness and outbursts of profanity. Such was the change in his personality and disposition that friends and former workmates could hardly recognise him. 'He was at times pertinaciously obstinate, yet capricious and vacillating, devising many plans of future operations which are no sooner arranged than they are abandoned . . . a child in his intellectual capacity and manifestations and yet with the animal passions of a strong man.'[4] 'Gage,' as his friends put it, 'was no longer Gage.'

Over the following years Gage worked at various occupations in various places, changing his jobs frequently and 'always finding something which did not suit him in every place he tried.' In 1854 he sailed for South America, where he worked as a stagecoach driver in Chile before returning to the USA in 1860 to live with his mother and sister in San Francisco. He died at 10 p.m. on 21 May 1861 following a series of seizures and convulsions. When, some five years later, Harlow finally learned of Gage's death he bitterly regretted that no post-mortem had been conducted whereby 'the precise condition of the encephalon at the time of his death might have been known'. In the vain hope of recovering some evidence regarding the precise injuries to Gage's brain from his, by now, well-decayed corpse, Harlow wrote to the young man's mother pleading to have his body disinterred so that his skull could be removed and retained for medical examination. He also asked that the tamping iron which, at Gage's request, had been buried beside him in the coffin, be recovered. Both these requests, he told his fellow doctors, had been granted 'with a magnanimity more than praiseworthy' for the benefit of medical science. At this point in his presentation Harlow, with the deftness of a magician pulling a rabbit from his hat, proudly displayed Gage's battered skull and his tamping iron to the assembled doctors.[5] In his paper Harlow had tentatively suggested that changes in Gage's personality might have resulted from the damage suffered to the frontal regions of his brain. It was a perceptive idea and one which, had it been taken up, might have advanced an understanding of brain function. But instead of praising Harlow for this insight the majority of doctors dismissed his idea out of hand. To understand the reason for their scorn we need to review the state of knowledge about brain science during the mid-19th century.

## The rise of the new brain science

At the time Harlow presented Gage's case history, the medical
consensus was that the brain functioned as a whole, with no area
being specialised for any particular task. It was a view doctors clung
to tenaciously despite the fact it had already been partly undermined
by two earlier discoveries. Seven years before Harlow wrote his
paper, a French surgeon named Paul Pierre Broca published an
account of a patient named Leborgne in the Bicêtre Hospital on
the outskirts of Paris. Over a 21-year period this man had gradually
lost the power of speech, although his comprehension and mental
functions remained unaffected. He was nicknamed 'Tan' by the
hospital staff since this was the only sound he could utter. After
Tan's death Broca performed an autopsy and found, as he had
expected, a lesion (damaged tissue) in the frontal lobe of the left
cerebral hemisphere. This led him to suggest that our ability to
speak depends on this small region of the brain, known today as
Broca's area.[6,7] Three years after Broca's discovery a German neurologist, Carl Wernicke, also used post-mortem findings to conclude
that a second area on the left side of the brain, located further back
than Broca's, was responsible for understanding the spoken word.

   While the medical profession readily accepted localisation for
both speech and comprehension of the spoken word, it chose to
reject Harlow's theory that personality and rationality could also
be localised with the brain. For one thing, both Broca and Wernicke
had based their ideas on post-mortem evidence. Phineas Gage had
taken the extent of his brain damage to the grave. For another,
the railway worker had never shown any overt signs of significant
brain damage. Following the accident his speech, movements and
memory all appeared unchanged.[8] It was British physiologist David
Ferrier who rescued both John Harlow and Phineas Gage from
obscurity. In 1878 he published a paper[9] dissenting from the

widespread rejection of Harlow's views. He argued that while the injury had spared those parts of Gage's brain responsible for movement and speech it had damaged a region known as the left prefrontal cortex, and that this damage probably explained what he described as Gage's 'mental degradation'.

## Modern-day Phineas Gages

One of the most intensively studied cases of long-term brain damage, similar to that suffered by Phineas Gage, was reported in 2001 by Maria Mataró and her colleagues at the University of Barcelona's Department of Psychiatry and Psychobiology.[10] Their 81-year-old male patient, identified as EVR, was born into a wealthy Barcelona family in 1916. When the Spanish Civil War broke out in 1936 he was a university student who soon became involved in the turbulent politics of the time. A warrant was issued for his arrest and police arrived at his second-floor apartment to seize him at gunpoint. EVR attempted to escape by climbing through a window and shinning down a drainpipe. The pipe gave way and he was hurled to the ground. The fall itself might not have caused serious injury but, unluckily for him, he landed head first on the iron spikes of a gate and was impaled through the head. The unfortunate young man remained conscious and even helped rescuers as they cut through the bar. In hospital the fragment of the spike protruding from both frontal bones was removed, the doctors noting that it had 'penetrated the left frontal region, passed through both frontal lobes injuring the left eyeball, and emerged from the right side'.

EVR survived his injury and was sufficiently recovered two years later to marry his childhood sweetheart and father two children. He went on to work in the small family firm until retirement. While his performance on basic mental tests remained normal, the researchers noted that his 'social, professional, and personal conduct was profoundly affected after the injury and was characterized by divorces, bankruptcy, and the inability to sustain normal work behaviour'. Throughout his long life he remained highly dependent on others. His job with the family firm mainly involved simple manual tasks that were always organized and checked by others, and even when carrying out routine domestic activities he required constant supervision. He was incapable of implementing plans or fulfilling responsibilities and had difficulties managing money.

'As a child, I realized that my father was a "protected" person,' his daughter recalled. 'When I was young I soon saw what the "problem" was, although I had always suspected it. At 17, I became part of this protection, and I still am.'[11] Restless and impatient at times, EVR was most notable for his apathy, lack of drive, and problems initiating, persisting with and completing tasks. On the plus side he was a relentlessly cheerful if somewhat tedious companion, recounting the same joke over and over again. Never irritable or hostile, he displayed no outbursts of rage nor showed any difficulty in controlling his emotions. This situation remained largely unchanged for the 60 years following his injury. 'To our knowledge,' comment the researchers, 'there are no descriptions in the literature of evolution of so long a period. This case illustrates that large frontal lesions can produce behavioural and personality changes . . . the protected and structured family and work environment has probably made it easier for him to lead a relatively normal life.'[12]

## What brain damage tells us about impulsivity

Studies like these help us understand how such injuries can lead to greater impulsivity as well as revealing a great deal about the neuroscience underlying the impulse. In both the cases described above injuries were caused, by the tamping rod and the gate spike, to a part of the brain known as the orbitofrontal cortex (the portion that lies in the front of the head above the eyes – often known as the OFC), with more damage being done to the left than the right side of the brain. Damage was also caused to a structure known as the anterior cingulate cortex (ACC). This resembles a 'collar' around the corpus callosum, the latter being a massive bundle of fibres that relays messages between the right and left hemispheres of the brain. The cingulate cortex plays a wide variety of roles, for example regulating blood pressure and heart rate, as well as enabling us to make decisions, express emotions, display empathy and anticipate rewards. Research also suggests that the ACC plays a key role in detecting and monitoring errors, evaluating the extent of such errors and suggesting what actions would be most appropriate to rectify them. Clearly this has huge implications for understanding the causes of impulsivity. If an impulsive person is unable to rapidly detect errors caused by their impulsive behaviours they cannot take timely steps to correct them.

Another brain region that helps distinguish between appropriate and inappropriate behaviours is a structure called the *nucleus accumbens*, which lies deep within what is, in evolutionary terms, the most ancient and primitive region of the brain. 'In order to learn which actions are the correct ones that eventually lead to reward, and which are not, some mechanism must "bridge" the delay between action and outcome,' explain Drs Rudolf Cardinal, John Parkinson and Barry Everitt. 'The [nucleus accumbens] is a reinforcement learning structure specialised for the difficult task of learning with, and choosing, delayed reinforcement.'[13] When this region is surgically removed in laboratory rats the animals become chronically impulsive, choosing small but immediate rewards over larger but more delayed ones. Until the nucleus accumbens matures, something that may not happen until after the age of 20 in humans, there is a disconnection between the amount of effort the teenager is prepared to expend on any activity and the level of reward expected. This helps to explain why, as a general rule, adolescents favour activities that combine the least amount of effort with the greatest and most immediate rewards. This is a topic to which we will be returning in Chapter 4 when I describe how the teenage brain matures.

Also affected by the waves of change in adolescence is the amygdala, whose name is derived from the Greek for 'almond' because of its shape. Together with the hypothalamus, the amygdala is responsible for learned fear and the 'startle response' as well as being involved in emotions such as anger and aggression, our sense of smell, and some types of learning and memory. The amygdala responds 'robustly' to wide-eyed expressions of both fear and surprise. This is due to the increases in exposure of the white sclera surrounding the iris, rather than the shape of the eyes themselves.[14] Changes to the amygdala may be responsible for the impulsive 'hot' versus 'cool' response of many teenagers as well as their propensity for mistaking the neutral or inquisitive facial expressions of others as indicating anger or aggression. Such misunderstandings can cause adolescents to view their world as being a more dangerous and hostile place than it appears to adults. This, again, is a topic that will be explored more fully in Chapter 4.

Drs Frank Benson and Dietrich Blumer[15] suggest that brain damage of the sort suffered by Phineas Gage and EVR are likely to produce two types of personality change. The first, characterised by poor planning, apathy, lack of drive and an absence of concern for the future, is associated with massive frontal lesions. The second, which they termed *pseudopsychopathy*, is 'best characterised by the lack of adult tact and restraints'. These impulsive behaviours were associated with damage to the orbitofrontal cortex (OFC) mentioned above. One of the least understood regions of the human brain, the human OFC is believed to regulate planning behaviour associated with sensitivity to reward and punishment. There is a potential link of OFC function to impulsive behaviour in that impulsive people are less sensitive to punishment and more sensitive to reward than non-impulsive people[16][17][18], approaching reward situations even when the risk of punishment makes restraint more appropriate.[19] Impulsive people show poor passive avoidance. That is, they continue doing something even though it brings them more pain than gain. This ties in with the suggestion that it is the frontal cortex that regulates and controls impulsive behaviour.

## Brain changes that change one's life

One case which illustrates this condition involves a 35-year-old man who underwent surgery to remove a brain tumour that entailed the removal of large portions of his frontal cortex. This resulted in profound personality changes. The man's marriage, previously happy and stable – he had fathered two well-adjusted children – ended in divorce. Within months he had embarked on a second, short-lived marriage. Previously a successful businessman with a keen financial sense, he entered into a series of failed business ventures. 'Although he is skilled and intelligent enough to hold a job, he cannot report to work promptly or regularly,' say Paul J. Eslinger and Antonio Damasio. 'He fails to take into account his long-term interests and often pursues peripheral interests of no value. Unlike his former self, he is a poor judge of character and is often inappropriate.'[20]

Dramatic accidents, such as those described above, are not, of course, the only ways in which people can become far more impulsive following some form of brain damage. The same outcome may follow injury through disease, drug-taking or age-related changes. Small lesions in the OFC can lead to a rare neurological condition known as *Witzelsucht*, from the German *witzeln*, meaning to joke or wisecrack, and *sucht*, meaning an addiction or yearning. *Witzelsucht* is characterised by an uncontrollable urge to tell inappropriate jokes, recount anecdotes that are pointless or irrelevant, and make puns. While onlookers find such behaviour tedious, embarrassing or inconvenient the patient considers them intensely funny. Elderly people are prone to this disorder because of the decreased mass of their grey tissue. The stereotype of the rambling, wisecracking grandfather may have its origins in this condition. Another bizarre consequence of damage to the frontal cortex is anarchic (or alien hand) syndrome. This causes sufferers to feel compelled to reach out and impulsively grasp nearby objects, or even other people, as if magnetically drawn to them.[21] In a related syndrome, known as utilisation behaviour, also linked to OFC damage, individuals are unable to resist the impulse to use any object within their reach, even when not needed.[22] If, for example, a pair of spectacles is placed on a table the patient may impulsively put them on. Should another pair be available he or she is likely to repeat the action, putting the second pair on top of the first.

It is clear that the regions of the brain described above, especially the orbitofrontal cortex and anterior cingulate cortex, play major roles in System I thinking. Not by making us more impulsive but rather by enabling us to exercise self-control over such impulses. By putting a 'brake' on our zombie brain they allow time for us to reflect on and possibly reconsider our thoughts, words and deeds. If these regions are damaged or destroyed by injury, age, illness or an accident, for example by a stroke or tumour, the loss of control may be irreversible. When weakened by alcohol or drugs they are less able, at least in the short term, to inhibit our impulsive behaviour. In both cases the zombie brain regains the power to determine our behaviour. This is not to suggest that everyone who behaves impulsively is suffering from some form of brain damage! While this may be true in a small minority of cases, there are many other factors, including genetics, brain chemistry, youth and life experiences, both in the womb and after birth, that lead to an increase in impulsivity. Which raises the question of why we have evolved with impulses that can give rise to so many potentially self-destructive and damaging behaviours. The answer lies, at least partly, in the way we strive to make sense of and survive in the world.

## Zombie brains and human survival

The next time you are in a crowded lift or a rush-hour train, take a surreptitious glance around. What you observe helps explain why impulses form our default mode of thought. When forced to share a limited space with total strangers, humans respond in exactly the same way as strange monkeys placed in the same cage. Surrounded by their unfamiliar fellow primates they send out non-verbal signals designed to defuse any aggression. They, like us, perceive such crowding as a potentially hostile situation. The anxiety this perception gives rise to causes them to withdraw into themselves and restrict any social contact to an absolute minimum. In packed lifts and congested railway carriages we behave in precisely the same way. We avoid making eye contact by averting our gaze, staring at the floor, or placing some object – a book, magazine or newspaper – between ourselves and our fellow travellers. We may resort to some form of displacement activity, such as fiddling with a pen, repeatedly glancing at our watch or, if in a lift, needlessly jabbing the control buttons. Only if the amount of time we are obliged to spend with strangers is extended for some reason – perhaps due to a signal failure when travelling by train – will the defensive barriers come down as people start reaching out, however tentatively at first, to their companions. Monkeys do the same. After studiously ignoring one another for a while they will start grooming as a means of demonstrating their position in the pecking order. Interestingly if they are Old World monkeys, that is those originating in Africa or Asia, the more dominant will be groomed by those lower in the hierarchy, while if they are New World monkeys, from South America, the more subservient will be groomed by the more dominant.

These rules of behaviour, in both humans and monkeys, have evolved over millions of years to enable both us and them to survive in a hostile world. Such survival frequently depends on drawing inferences and making decisions very rapidly and with sufficient accuracy to deal successfully with whatever situation confronts us. This process typically involves zombie brain System I thinking which enables us to assess any new situation in this way. Within a few hundredths of a second of meeting someone for the first time we usually get a sense of whether we like them, are sexually attracted to them, are indifferent to them or are uncomfortable in their presence. The feeling of not liking someone (System I thinking) without being able to explain why (System R thinking) was neatly summed up by a 17th-century student named Thomas Brown in a verse dedicated to his university dean Dr Thomas Fell:

> I do not like thee, Doctor Fell,
> The reason why I cannot tell;
> But this I know and know full well,
> I do not like thee Doctor Fell.[23]

To meet such potentially dangerous situations the brain evolved a large number of 'rules of thought' otherwise known as *heuristics*. Based on the Greek word for 'find' or 'discover', heuristics are simple, efficient rules that come into action rapidly and automatically whenever the need arises. 'Different environments can have different specific fast and frugal heuristics that exploit their particular information structure to make adaptive decisions,'[24] say researchers Gerd Gigerenzer and Peter Todd, who use the phrase 'adaptive toolbox' to describe the hundreds of heuristics the zombie brain uses to negotiate through life. While many heuristics are hard-wired into the brain and reflect the survival demands of our earliest ancestors, most are, however, handed down from one generation to another and one culture to the next. Surrounded by the trappings of modern life, with its ever more sophisticated forms of travel, space exploration, soaring skyscrapers, iPads and social networks, it is all too easy to forget just how recent is everything we regard as civilised. While tool-making hominids, the ancestors of modern man, emerged some three million years ago, the early foundations of today's civilisation were only laid down some 12,000 years ago, at the end of the last ice age. For 99.5 per cent of the time our species has existed on earth human behaviour was dictated almost entirely by System I thinking. During this period change occurred with glacial slowness. In an era where 100,000 years might pass before a new technique developed, the well-honed if limited heuristics on which the zombie brain relied proved perfectly effective. As author Ronald Wright points out in *A Short History of Progress*, 'No city or monument is much more than 5,000 years old. Only about seventy lifetimes, of seventy years, have been lived end to end since civilisation began. Its entire run occupies a mere 0.2 per cent of the two and a half million years since our first ancestors sharpened stone.'[25]

When humans were evolving, therefore, so many physical threats to survival existed that the brain evolved heuristics to enable extremely rapid decisions to be made and actions taken. The startle response, for example, which instantly transforms a state of relaxation into one of maximum alertness, could mean the difference between life and death when danger lurked behind every bush. Our problem is that while they proved effective thousands of years ago – I would not be here to write these words and you would not be here to read them had that not been the case – today they can prove totally and embarrassingly inadequate and potentially fatal for human survival. Take for example the growth in new information, a growth that drives exponential change. It has been calculated that from the dawn of modern civilisation to 2003, a period of around 5,000 years, human activity generated around five exabytes (that is five billion gigabytes) of information. Between 2003 and 2010 the same amount of information was generated every two days. By 2013 it was generated every ten minutes. Within hours it had exceeded all the information contained in all the books ever written![26] Yet, to continue the computer analogy, we are still attempting to make sense of the 21st-century world using hardware and software that received its last upgrade more than 50,000 years ago.

The heuristics that drive our zombie brain, and which reveal themselves in the mindless actions and impulsive actions they generate, are typically triggered by features in our environment, for example sights, sounds, or smells whose influence occurs below the level of conscious awareness. As a result we often say and do things which take even us by surprise and for which we can offer no rational explanation. Later on in the book I will describe some of these subliminal influences and explain why they cause us to act in impulsive and sometimes self-destructive ways.

# CHAPTER 4

# The Teenage Brain –
# A Work in Progress

'Human life is a long-term, high investment, knowledge
economy, and it is teenagers who made it that way. This is
why adolescence is not an irritating transitional phase, but
is literally pivotal in the human life-plan: it is the fulcrum
about which the rest of our life turns.' David Bainbridge,
*Teenagers: A Natural History*

On 28 July 1999, Kathleen Grossett-Tate, a Florida highway patrol
officer and single mother, agreed to look after Tiffany Eunick, the
6-year-old daughter of her best friend. Although Kathleen's
12-year-old son, Lionel, had only known the little girl for a few
weeks, they appeared to get along well together.  After supper,
Tiffany and Lionel went into the living room to watch TV while
Kathleen completed some household chores upstairs. At around
ten o'clock she heard a lot of noise in the living room and yelled
at the children to be quiet. Forty minutes later her son came up
stairs and calmly announced that Tiffany was dead. Lionel, a stocky
pre-teen who weighed 12 st 2 lbs, told his mother they had been
copying wrestlers seen on TV when the slightly built little girl's
head had struck the table. At first it seemed a tragic accident caused
by childish rough-and-tumble play gone disastrously wrong. The
pathologist's report, however, told a different story. It described
how, far from the single fall Lionel had described, the six-year-old
had suffered a severe beating lasting as long as five minutes. Her

35 injuries included a ruptured spleen, partially detached liver, fractured skull, brain contusions, damage to her ribs, lacerations and bruises all over her small body.

During the boy's trial for murder, the court was told that from an early age Lionel had been in constant trouble at school. He had been suspended on 15 occasions for fighting, lying, assault and theft. His IQ was assessed at 90 and his 'social maturity' evaluated as being equal to that of a 6-year-old. Dr Michael Brannon, a forensic psychologist at the Institute for Behavioural Sciences and the Law, who had examined the pre-teen, told the court that while not mentally ill, the 12-year-old possessed 'a high potential for violence, uncontrolled feelings of anger, resentment, and poor impulse control'. It took the jury less than two hours to return a guilty verdict. Lionel, who had taken little interest in the trial and spent most of the time drawing pictures at his lawyer's table, was sentenced to life imprisonment without parole. He was the youngest person in US history to be handed down such a sentence.

In late 2003 a Florida appeals court upheld a defence submission that the judge had been in error by refusing to test Lionel Tate's mental competency to stand trial. They said that in the light of Tate's age, his unfamiliarity with the legal system and the complexity of the prosecution's case the trial judge had been wrong to assume he was competent to stand trial. A few weeks later, by now a teen-ager, Lionel was released into his mother's care. In exchange for a guilty plea to second-degree murder, his sentence had been reduced to one year's house arrest and ten years' probation. In addition, he was to receive mandatory counselling and undertake community service. At first Lionel's life seemed to be returning to normal. But his story was not to have a happy ending. As time passed he lost touch with his supporters and failed to attend the church groups which had helped him while he was in prison. In September the following year, police arrested him for carrying a knife. His probation was extended to 15 years and he was warned that another violation would send him back to jail. Seven months later he was arrested for a third time and accused of armed robbery. The court heard how, after ordering some pizza, Tate had held up the deliveryman with a revolver and then forced his way into a neighbour's flat to hide from the police. On 19 February 2008, after various hearings, Tate was sentenced to ten years in state prison to run concurrently with a 30-year sentence for violating his probation.

Lionel Tate, clearly, had little or no way of controlling his violent and impulsive behaviour. But, although the levels of violence he displayed are exceptional, an inability to keep impulses in check is a natural and normal part of growing up. As anyone who has dealings with young people on a regular basis will readily confirm, adolescence is a period of development characterised by impulsive, and often risk-taking, behaviour. This desire to test boundaries and rebel against restrictions arises from evolutionary pressures to move from dependence to independence.[1] Unfortunately the consequences – which include alcohol and drug abuse, juvenile crime and sexual experimentation – can sometimes prove tragic[2] and often, as in the case of Lionel Tate, have long-term consequences.[3]

## Impulses in infancy

Impulsive behaviour is by no means confined to the teenage years. It can be observed even among pre-schoolers. Some 30 years ago, I spent 12 months in a playgroup studying the body language of very young children – infants for whom non-verbal rather than verbal communication was their chief way of interacting both with one another and with adults. Using one of the earliest video

recorders, a cumbersome black-and-white camera attached to a
heavy recording device the size of a small suitcase, I videotaped
and later analysed their interactions.[4] It was quickly apparent to
me that, even at 2 years old, levels of impulsivity varied enor-
mously.[5] Some youngsters were active, dominant and highly impulsive. If they wanted another child's toy they simply grabbed it. If
they wanted to use the slide others would be roughly elbowed out
of the way. Such infants were seldom deterred by the presence of
a nearby adult, nor did reprimands cause them to modify their
impulses.

On the basis of such emerging motor competences, it is possible
to distinguish between varying levels of impulsivity within the
first few months of life.[6] One study found evidence that those
infants whose temperament was judged 'difficult' when just 6
months old would be regarded as more than usually impulsive on
starting school.[7] However, impulsivity at the age of 2 or 3 does not
necessarily mean that a child will remain that way as he or she
grows up. Indeed, many other researchers have found no correla-
tion between measures of impulsivity during the first 12 months
of life and impulsive behaviours when older. 'This seems reasonable given that the first year of life is a period of rapid development, high plasticity, and restricted behavioural capability, which
may mask the condition of the brain,' comment Drs Jorge Daruna
of Tulane University School of Medicine and Patricia Barnes of
New Orleans Adolescent Hospital.[8]

While studies of twins separated at birth have shown impulsive
behaviour to have a significant genetic component, a tendency to
act without thinking is by no means due to nature rather than
nurture.[9] Daruna and Barnes also make the point that early indications of impulsivity often reflect only 'a transient lack of
synchrony in the development of a relative neural system'. It may
also be that the approach of parents and the infant's social

environment are capable of 'attenuating the bias towards impulsive behaviour'.[10] Mothers who encourage independence are less punitive and able to talk with their child warmly and frequently during the pre-school years and are less likely to have impulsive children. Overly controlling, intrusive mothers, by contrast, are more likely to encourage any genetic tendencies towards high impulsivity.[11]

The baby's development while still in the womb may also play a role in creating impulsive children. External factors, such as the mother's stress, illness, or obesity, which affect the intrauterine environment during pregnancy, can profoundly influence brain development and the ways in which the genetic blueprint is expressed.[12]

Poverty, too, can have far-reaching consequences through the infant's heightened vulnerability to common medical illnesses across the course of a lifetime. Children raised in families with low socio-economic status have been found to suffer from elevated rates of infectious, respiratory, metabolic, and cardiovascular diseases in adulthood, independent of traditional risk factors for those conditions. In addition, the psychological impact of poverty can also influence levels of impulsivity, impetuousness and reckless behaviours.[13]

The good news is that these consequences are not inevitable. Indeed, studies have shown that around half the children born into poverty live long and perfectly healthy lives. What makes the difference is the amount of love and attention provided by the child's parents or other caregivers who are able to imbue children with the sense that the world is a safe place and other people can be trusted. 'These beliefs may enable disadvantaged youngsters to read less threat into their social worlds,' says Dr Gregory Miller of the University of British Columbia, 'with a consequent reduction in the wear and tear that such vigilance can place on bodily systems . . . [They] also help children to learn emotion-regulation strategies, so that when they do encounter stress, the physiological consequences are attenuated.' From a biological perspective these benefits may result from increased production of oxytocin, a peptide released when people experience warmth and security. Animal studies indicate that oxytocin counteracts some pathogens involved in ill health.

## ADHD – when impulses take over

Impulsive behaviour is a key feature of ADHD (attention-deficit hyperactivity disorder), a condition affecting about one in twenty children – mainly boys – worldwide.[14] In 40 to 60 per cent of cases ADHD persists into adolescence and adulthood, leading to problems ranging from poor academic performance and socialisation to increased traffic accidents.[15] The most widely used method of treatment is to prescribe drugs, such as methylphenidate, atomoxetine and dexamfetamine, in an attempt to bring the condition under control. Unfortunately, around one in five ADHD sufferers fail to respond to drugs[16] while in many other cases the response is only partial. Furthermore, all drugs have side effects[17][18]; they can also be habit-forming and open to abuse. Long-term follow-ups have found that when children stopped taking the drug their clinical symptoms of ADHD reappeared.[19][20]

Problems such as these have led some therapists, mainly in the USA, to start using a relatively new form of treatment known as neurofeedback training. This involves teaching sufferers how to inhibit their impulses by learning to control their 'brain waves' by playing computer games via sensors attached to their head. These detect different patterns of electrical activity and use the signals to control characters on a screen. Although my laboratory does not undertake therapy as such – we refer all enquiries in the UK to one of the relatively few professional neurofeedback trainers – we have undertaken research into the efficacy of this approach. The results certainly appear promising.

In one such session, two teenage boys play a game involving a race between a red and a blue caterpillar. Thin wires snake down from electrodes pasted onto their scalps to a control box on the bench before them. This detects electrical activity in the brain and uses these 'brain waves' to move the caterpillars across the screen. By learning how to put themselves into a particular frame of mind the youngsters are able to get their caterpillars to move at speed across the screen. Mark, one of the two boys, has been diagnosed with ADHD. His symptoms include impulsivity, an inability to concentrate in class and hyperactivity. Ryan, his classmate and companion in the study, exhibits no such symptoms. During the game, Ryan's blue caterpillar moves rapidly down the track as he easily manages to reduce slow-moving brain waves, known as 'theta waves', while at the same time increasing faster-moving 'beta waves'. Because he is unable to focus his attention on the game, Mark produces more theta and fewer beta waves. As a result his red caterpillar barely moves off the start line. Over a period of time, however, he teaches himself how to reduce his theta and boost his beta waves. By doing so he brings his ADHD under control. Such training, which can take from a few weeks to several months, has been shown to significantly reduce the symptoms of the condition in up to 40 per cent of children.

## How the child's brain develops

Although some adults still dismiss ADHD as 'naughtiness' dressed up in a fancy medical diagnosis, most doctors agree it is a neuro-developmental psychiatric disorder resulting from a failure of the

brain to evolve adequate inhibitory controls. Throughout childhood, and especially between the onset of puberty and early adulthood, the brain undergoes constant and significant changes. During
the early years the number of neurons (nerve cells) in the brain
steadily increases as do the connections between them. From the
age of around 11 in girls and 13 in boys, however, these nerve cells
begin to be 'pruned' so as to produce a more efficient brain. 'The
brain can grow extra connections sort of like branches, twigs and
roots to use a gardening metaphor and then after it has these
connections there's also another gardening metaphor called pruning
or cutting back or eliminating the excess or unused connections,'
explains Dr Jay Giedd who heads the National Institute of Mental
Health's Longitudinal Brain Imaging Project, 'it's this process of
overproducing and then having fierce competition amongst all
these connections to see which ones are most useful and which are
most helpful for us to adapt to the environment.'[21] Neuroscientist
Professor Stephen Wood, from the University of Melbourne, has
likened this process to the creation of a sculpture. 'You end up with
less stone or clay, but it is a better finished work,' he says.[22]

The introduction in the late 1980s of magnetic resonance
imaging, a non-invasive technology that enables the living brain
to be studied in exquisite detail, caused views about its development to be drastically revised. Between 1989 and 2007, Jay Giedd
and his team scanned the brains of some 2,000 people at approximately two-year intervals. In addition to making around 5,000
scans they also conducted neuropsychological and behavioural
assessments and collected samples of DNA. Of these 2,000, almost
400 aged between 3 and 30 served as the models for typical, healthy
brain development. 'Adolescence is a time of substantial neuro-
biological and behavioural change, but the teen brain is not a
broken or defective adult brain,' says Dr Giedd. 'The adaptive
potential of the overproduction/selective elimination process,

increased connectivity and integration of disparate brain functions, changing reward systems and frontal/limbic balance, and the accompanying behaviours of separation from family of origin, increased risk taking, and increased sensation seeking have been highly adaptive in our past and may be so in our future. These changes and the enormous plasticity of the teen brain make adolescence a time of great risk and great opportunity.'[23]

Changes to the brain during puberty start in a region at the back of the brain called the cerebellum, or 'little brain'. Its function is to control physical coordination and the processing of sensory information. When boxers become 'punch drunk' as a result of blows to the head, it is this area of the brain that shows the most damage. Next to mature is the nucleus accumbens, followed by the amygdala and finally the prefrontal cortex, located behind the forehead and superior to the orbitofrontal cortex discussed at length in the previous chapter. In general, the process of maturation involves an increase in connectivity and integration, with previously distributed regions becoming more closely associated. The amount of grey matter *decreases* from the pre-teen peak while white matter, comprising a fatty substance called myelin, *increases*. Wrapped around most of the nerve fibres, myelin acts like an electrical insulation speeding the flow of signals and turning the brain into a super-fast information highway. One writer has compared it to 'switching from dial-up internet to broadband'.[24] There is also an increase in the size of the corpus callosum. This structure, comprising some 200 million fibres, connects the right and left hemispheres and forms one of the major communication channels between the two sides of the brain. Finally, there is a change in the balance between the emotion-generating limbic system and the higher regions of thought in the frontal cortex. So what effect do these progressive changes have on the way adolescents make sense of, and respond to, the world around them?

The fact that the frontal parts of the brain, responsible for self-control, judgement and caution, are the last to mature partly explains teenagers' reputation for impulsivity, recklessness and emotional volatility. An immature prefrontal cortex is one of the reasons why teenagers show poor judgement and too often speak and act impulsively. As these changes are occurring in their brain, surging hormones are driving teenagers, especially boys, to seek out new thrills and experiences. 'There is a developmental mismatch, with increased drive and no brakes,' says Professor Stephen Wood.

## Mental illness and the teenage brain

The teenage brain is particularly vulnerable to various forms of mental illness. The origins of around three-quarters of all mental illnesses occur between the ages of 15 and 25, even though the diagnosis may not be made until much later. A report in 2005 estimated that some 2.7 million children and adolescents are suffering from severe emotional or behavioural difficulties. These may persist into adulthood and lead to lifelong disability, including more serious forms of mental illnesses. Research by Stephen Wood and his colleagues has shown that the brains of young people who fall ill in this way seem to develop faster than normal. There is also a greater loss of grey matter, for their age, among the mentally ill than is found in those who remain healthy. 'The implication of our growing knowledge of brain-behaviour mechanisms of adolescent conditions should provide insights into the risk of particular adolescents for morbidity and mortality,' comments Dr Elizabeth McAnarney. 'Preliminary data are promising so that as we begin to understand the complexity of and specificity of each of these conditions, we shall be able to diagnose and treat conditions earlier.'[25]

# Is childhood impulsivity all down to biology?

Laurence Steinberg, of Temple University, Philadelphia, has spent a lifetime studying the adolescent brain and discovering some of the root causes of youthful impulsivity.[26] His research suggests that adolescent impulsivity arises from an imbalance between sensation seeking, which increases dramatically after puberty, and self-control that does not fully mature until the early to mid-twenties. As a result of this breakdown in communication between System I and System R thinking teenagers are not only more likely to act impulsively but may also fall into the trap of deliberating too long over problems better dealt with using gut feelings. In one study teenagers were asked whether some obviously dangerous antics, such as setting light to one's hair, were 'good ideas'. They took significantly longer than adults to answer and made fewer demands on brain regions concerned with control. Such a period of reflection was not found when the activities were perfectly safe – 'eating a salad', for example. The result, as Steinberg puts it, is that 'the accelerator is activated before a good braking system is in place'[27], with middle teen years – 14 to 17 – being an especially vulnerable period when this disparity between the two systems is largest. He found that teens spend less time than young adults in thinking before acting and show a stronger preference for immediate than postponed rewards. In other words they are significantly more under the control of System I thinking and their zombie brain than are adults.

Peer-group pressure also exerts a far stronger influence on teen-agers than on young adults. Steinberg had teenagers and adults aged between 19 and 22 perform a simulated driving task in which they approached a set of traffic lights on amber. They made this drive either by themselves or while being observed by two same-sex, same-age friends. When being watched, teenagers were far more

likely to run the yellow light and risk a crash.[28] While the same effect
was also found with the young adults, it occurred to a far lesser
extent.[29] Findings such as this help explain why so much risky adolescent behaviour, such as drinking, reckless driving, or delinquency,
occurs in groups. Young people spend a lot of time in the company
of their peers, and the mere presence of other adolescents increases
the rewarding aspects of risky situations by activating the same
neural pathways that are activated by non-social rewards when alone.
This is a topic I will return to in Chapter 11 when I consider copycat
behaviour in riots and violent public disorder.

The popular stereotype portrays teenagers as impulsive and irrational. As young men and, to a lesser extent, young women who
are often reckless and irresponsible. As people who drink to excess,
experiment with drugs, start fights, drive dangerously, commit petty
crime and acts of violence and engage in unprotected sex whenever
the opportunity arises! In the minds of most tabloid editors and
many of their readers they are, to employ a term used by the pros-
ecutor at Lionel Tate's trial, 'juvenile super-predators'. This biased
view of adolescence is contradicted by vast numbers of detailed
and reliable research studies. Indeed, the logical reasoning abilities
of 15-year-olds have been found comparable to those of adults.
Adolescents are also as competent as adults when it comes to
perceiving risk and estimating their own vulnerability. So while
System I thinking can (and all too often does) dictate their actions,
their perceptions and the assumptions they make, the zombie brain
is perfectly capable of being brought under control.

## Risk and reward in the teenage years

The teenage years are undoubtedly a time of risk. But they are
also one of opportunity, creativity and learning. Because the

maturing brain is still 'plastic', adolescence is a period when firm but sensitive care from adults who respect and have high expectations of the youngster can transform their life. The way in which such help and advice is given has a significant bearing on how it is received and acted upon. While providing adolescents with reliable information about the risks of unprotected sex, substance abuse or reckless driving usually improves the way they think about such activities, it seldom changes their actual behaviour. Research shows that, in general, reductions in health-compromising behaviours are most likely to occur when changes are made to the contexts in which those risks are taken. For example, increases in the price of a packet of cigarettes, enforcement of the licensing laws or free contraception can all change risky behaviour for the better. 'We shouldn't be criticising teenagers, we should be celebrating them,' asserts David Bainbridge in *Teenagers: A Natural History.* 'We've become blind to the fact that our teenage years are the most dramatic, intense and exciting of our lives. Everything is very vivid.'[30] During the seven teenage years, the course is set that will take adolescents on their voyage into adult life. On the decisions they make during this confusing and chaotic period of their lives, when brain and body undergo vast changes, may well depend their entire future.

# CHAPTER 5

# Impulse and the Senses

'All we have to believe with is our senses, the tools we use
to perceive the world: our sight, our touch, our memory.
If they lie to us, then nothing can be trusted.'
Neil Gaiman, *American Gods*

In 1572, a royal ball was held in the Louvre to celebrate the
marriage of Henri de Bourbon, Prince of Condé, to 16-year-old
Marie de Clèves, a young woman possessing 'great beauty and
sweetness of nature'. Recounting the events of that evening some
two centuries later, the French anatomist Jules Germain Cloquet
wrote: 'After dancing for a long time and feeling slightly overcome by the heat of the ballroom, Marie went into a dressing
room where one of the Queen's maids helped her to change
into a clean chemise. The Duke of Anjou [who later became
Henry III] coming by chance into the dressing room, happened
to pick up the discarded chemise and used it to wipe his face.
From that moment on, the Prince conceived the most violent
passion for her.' While the resulting scandal outraged the court
and shocked French society, it offers an excellent illustration of
how powerful impulses can be triggered in subtle and unexpected
ways by stimulating one or more of our senses below the level
of conscious awareness. By senses, I mean groups of sensory
cells that respond to a specific physical phenomenon and corres-
pond to particular brain regions where signals are received and
interpreted.

# Twenty-one ways to make sense of the world

We humans possess around 21 different senses, rather than the five
– seeing, hearing, tasting, touching and smelling – normally recognised. These include:

**Sight**: Technically this is composed of two senses, based on the
fact that our eyes have receptors (cones) that produce colour vision
and others (rods) that enable us to see in dim light.
**Hearing**: Hairs in the inner ear are sensitive to changing pressure
in air or water and communicate these vibrations to specific audi-
tory centres in the brain.
**Smell**: Communicated by chemically sensitive cells in the olfactory
epithelium, this is one of the most primitive and influential of
our senses and one I will be exploring in much greater detail a
little later in this chapter.
**Taste**: This is considered by some researchers to be five senses in
one, since the tongue has receptors specialised to detect things that
are sweet, salty, sour, bitter, and *umami*. The latter is a Japanese
word coined in 1908 to denote a pleasant, meaty taste; its receptors
detect glutamate, an amino acid found in meat and some artificial
flavouring. Incidentally, the 'taste map' showing areas of the tongue
supposedly dedicated to specific tastes is completely wrong and
based on a misunderstanding of some research conducted in
Germany during the last century.
**Touch**: This sense is distinct from those that send information
about pressure, temperature, pain, and even itch to the brain.
**Thermoception**: Our ability to feel heat and cold is also regarded
as the work of more than one sense. Not only are there two separate detectors for heat and cold, but a third, entirely different type
of thermoceptor is located in the brain and used to monitor
internal body temperature.

Other senses include **proprioception**, which lets us know where parts of our body are located in respect to other body parts; **tension sensors** that monitor muscle tension; **nociception**, which communicates the sensation of pain; **equilibrioception**, which enables us to keep our balance; **stretch receptors**, located in our lungs, bladder, stomach, skeletal muscles, and gastrointestinal tract; **chemoreceptors**, involved in detecting hormones and drugs in the bloodstream; and even **magnetoception**, an ability to detect magnetic fields. All these are in addition to senses that alert us when we feel hungry or thirsty. While triggering any one of these senses can result in impulsive behaviour, here I shall be discussing three of them, those of smell, sound and thermoception, because these three are most likely to be manipulated by others for commercial purposes. The vital role played by our visual sense in triggering impulses will be considered in the next chapter.

## The impulsive power of smell

Aside from the influence of genetic factors on body odour, which helps to explain why members of the same family not only look but smell alike, we all possess an entirely personal 'smell signature'. It is this unique smell that enables bloodhounds to track specific individuals, even in a crowded urban environment. Our 'smell signature' comprises a blend of skin, hair and glandular secretions, food and drink choices, odours in our surroundings, and any perfume, after-shave or deodorant we use.[1] But the most distinctive odour of all tends to be sweat. On its own this is capable of triggering impulsive sexual desire. The German sexologist Richard Kraft-Ebbing, for example, described a 'voluptuous young peasant man' who boasted how he had 'seduced quite a considerable number of chaste girls without difficulty by wiping his armpits with his handkerchief while dancing, and then using this handkerchief to wipe the face of his dancing partner'.[2] Although most human sweat is odourless and serves to regulate body temperature, specialised apocrine glands – located in the armpits and around the genitals – produce a secretion which contains part of the cell itself. These glands are triggered by emotional responses, including fear, anger and sexual arousal. The important role of apocrine glands in sexual desire and behaviour is indicated by the fact that they start functioning around puberty, peak with sexual maturity and become less active as people age.[3]

We develop our powers of odour discrimination at a very early age, as studies by Aidan MacFarlane of the Department of Experimental Psychology at Oxford University have shown.[4] Mothers taking part in his research were asked not to wash or put lanolin ointment on their nipples after feeds and to wear standard gauze breast pads. One soiled and one clean pad were then placed on a support in such a way that each touched the cheek of a baby, which could turn its head by 45 degrees to smell either of the pads. Using this simple device, MacFarlane demonstrated that infants as young as 10 days old are not only far more attracted by their mother's breast odours but can easily distinguish them from those of strangers.

But why should sweat, which many people consider more a passion killer than a passion stoker, have such a powerful effect? The answer lies in pheromones, contained in the sweat, that exert their seductive influence below our level of conscious awareness. The term pheromone is derived from two Greek words, *pherein* meaning 'to transfer' and *hormōn* meaning 'excite' or 'urge on'. First identified by Peter Karlson and Martin Lüscher in 1959, pheromones are molecules that animals use to communicate with one another. Aristotle described how some butterflies and moths were attracted to one another on the basis of odours, observations

confirmed during the 19th century by the great French naturalist
Jean Henri Fabre.[5] The effects of pheromones on humans were
first described in the early 1950s by the French physiologist Jacques
Le Magnen. His interest was sparked, so the story goes, by the
sensitivity of his female assistants to exaltolide, a chemical used
in perfume manufacture which has a heavy musk-like odour.
Whenever they approached a flask containing this substance the
women became highly aroused and, in some cases, inexplicably
upset. Male colleagues working alongside them were unaffected
and, in most cases, unable even to detect the smell. Equally
surprising was the fact that sensitivity to the exaltolide varied with
the menstrual cycle. Subsequent research showed that sensitivity
is at a maximum during ovulation, a fact which enables some
women to use it to make an accurate estimation of when they are
ovulating.

## Nerve Zero and our sense of smell

Our sensitivity to pheromones is (probably) due to Nerve Zero
or the terminal nerve (nervus terminalis)[6], first identified in
sharks in 1878, by the German scientist Gustav Fritsch, and in
humans some 35 years later.[7] However, the precise function of
this little-known nerve remains a matter of conjecture to this
day. This is one reason, perhaps, for its absence from most
anatomical textbooks; another is that, being extremely thin, the
nerve tract is often destroyed when the brain is removed from
the skull during a post-mortem. From the nose, Nerve Zero
enters the brain alongside the olfactory nerve (cranial nerve 1)
responsible for our sense of smell. It does not, however, terminate
at the olfactory bulb – the part of the brain where smells are
analysed – but at areas involved in the regulation of reproductive

behaviour. This has led some scientists to believe that it plays a crucial role in reducing the risk of someone falling in love and having sex with close relatives. By detecting pheromones, Nerve Zero identifies those possessing a similar genetic make-up. This, at least, is the conclusion drawn from studying an isolated religious community in the USA called the Hutterites.

## What Hutterites reveal about the sexual impulse

Between 1874 and 1879 a group of Anabaptists, known as Hutterites, fled religious persecution in Russia and ended up in the USA. Four hundred of these had developed a communal form of living based on the Acts of the Apostles. Today more than 350 clans, the ancestors of these 19th-century settlers, extend through the Dakotas, Minnesota and western Canada.[8] Because marriages take place within a limited gene pool, the Hutterites should experience significantly increased risks of spontaneous abortions and abnormalities among their babies.[9] In the event, the vast majority of their women give birth to perfectly normal and healthy infants. This fact puzzled geneticists who noted: 'Given the close genetic relatedness between individuals from within the same clan it comes as a shock that deleterious recessive traits don't run rampant and, rather, that healthy babies are the norm.'[10]

For Dr Rachel Herz, one of the world's leading experts on the psychology of smell, it raised the question of how, with so limited a choice of marriage partners, 'inbreeding' is avoided. The answer lies in the fact that mate selection among the Hutterite young is far from random.[11] Rather, they fall in love with partners whose genetic make-up is as different as possible from their own. They do so unconsciously and thanks to their sense of smell.

Studies suggest that one of the tasks of Nerve Zero is to identify incoming macromolecules, called the major histocompatibility complex (MHC), comprising hundreds of different proteins. Known, in humans, as the human leukocyte antigen (HLA), they help to regulate the immune system and produce an odour unique to every individual. 'The more variability in your HLA, the better your immune system,' explains science writer Kayt Sukel. 'It was long hypothesised that folks were most attracted to those who had an MHC complex as different from their own as possible – a little sniffable yet unconscious note to lock on to the mate that will help us produce the strongest, healthiest offspring.'[12] When two sets of sufficiently different MHC genes are detected by Nerve Zero, chemicals triggering increased sexual arousal and desire are released to make pair bonding and mating more likely. When similarity is detected, however, both arousal and desire are turned off by the same mechanism. This reduces the chances of someone accidentally mating with a close relative, since they are likely to share the same variety of MHC genes. By avoiding such partners they not only enhance their baby's resistance to disease after birth but also help ensure the pregnancy reaches full term.[13][14]

Claus Wedekind and his colleagues at the University of Berne, Switzerland, put this to the test by asking young male students to wear cotton T-shirts at night. Female students with the same average age were each asked to rate the odours of six T-shirts, three belonging to men whose MHC profile was dissimilar to their own and three from males whose MHC profile was similar. Women whose MHC profile was dissimilar to that of the man considered his odour more agreeable than did women whose MHC profile was similar. This difference was, however, reversed when the women were taking oral contraceptives. Under these circumstances the man's body odour was perceived as being more pleasant by women whose MHC was similar. 'Our findings show that some genetically determined odour components can be important in mate choice,' says Claus Wedekind. 'The observed mate preference could be a means to efficiently react to pathogen pressures. If so, the negative consequences of disturbing this mechanism, by the use of perfumes and deodorants or by the use of the contraceptive pill during mate choice, need to be known by users.'[15] While the idea that even undetectable odours may play a central role in sexual desire and our choice of partners may not be very romantic, it explains why people fall so helplessly in love, behave so strangely while in love and suffer such misery if their relationship comes to an end.

## Impulses and subliminal smells

Although less research has been done on subliminal smells than images, there is good evidence to show that the former may be far more powerful in influencing impulsive behaviours than was previously realised. In a study exploring the role of subliminal smells on liking in general, participants were asked to smell a citrus compound, rated as pleasant; anisole (a smell similar to aniseed), rated as neutral; and valeric acid, which most people consider extremely unpleasant. Although the odours were so diluted as to be unidentifiable, they still influenced the extent to which people were liked or disliked. In the presence of a pleasant odour people were impulsively regarded as more likeable than when viewed with an unpleasant odour being present. Heart rate slowed in the presence of a pleasant odour and increased with an unpleasant one. 'The fact that minute amounts of undetectable odorants could elicit salient psychological and physiological changes highlights the acute sensitivity of human olfaction, which tends to be underappreciated,' says Dr Wen Li from the University of Wisconsin–Madison. 'In particular, the reliable effects of the subliminal unpleasant odour imply that there may be a specialised high-affinity sensory channel for odours carrying threatening messages.'[16]

Another study, conducted by Dr Joshua Tybur and his colleagues at the University of New Mexico, examined the influence of odours on sexual behaviour. In this, male volunteers were asked to fill in a questionnaire about their use of condoms. The task completed, they were instructed to go and refresh themselves at a nearby drinking fountain. While they were away, one wall of the room was sprayed with a tiny amount of a joke-shop odour smelling of faeces. A second group of men, matched for age and educational background with the first, completed the same set of questions while being exposed to the faint but foul-smelling faecal pong. When questioned about their intention to practice safe sex in future, those in the group exposed, without their knowledge, to the disgusting odour were significantly more likely to say they would use condoms than those who had not. The psychologists' explanation was that the odour, although barely detectable, had been sufficient to trigger fears of disease. This anxiety increased their desire to safeguard their health, so making them more receptive to condom use.[17]

## Perfumes and passions

According to Greek legend, after women on the island of Lemnos refused to pay her tributes, the goddess Aphrodite cursed them with such a horrid stench that their husbands left them for other women.[18] Today the same fear of offensive body odour and a desire to make oneself sexually attractive and desirable are two main reasons why the manufacture and sale of perfumes and cosmetics has become a £170-billion-a-year global industry. While some find the heady aroma of unwashed bodies sexually alluring – Napoleon famously wrote to Josephine instructing her not to bathe before he paid a visit – for most people an overly ripe body odour is completely unacceptable. This is one reason, perhaps, why wealthy Romans indulged in aromatic baths as a regular prelude to love-making, enjoyed being massaged with sweet-smelling unguents and used copious amounts of perfumes on their head, body, hair and clothing with aromatic spices to sweeten their breath. Civet and ambergris were especially popular among the leisured classes of imperial Rome, and given the importance of vanilla as an aphrodisiac it is interesting to note that the name of this spice is a diminutive of the Latin word *vagina*. When, during the 16th century, Sheikh al-Nefzawi of Tunisia wrote one of the world's earliest sex manuals, it was no coincidence he titled it *The Perfumed Garden*.

One substance used in modern male grooming products is androstenedol, a chemical structurally related to testosterone, the hormone produced in the testes which is found in male sweat and urine. Its purpose is to make women swoon at the wearer's feet and research suggests that it may indeed partly do so. Dr David Benton of University College, Swansea, a leading researcher in this field, found that females reacted more strongly to the hormone during the middle of their menstrual cycle, when it caused them to assess their own mood as more submissive. But he adds a note of caution: 'Sexual attraction relies on many factors including personality, social skills, past experiences and the present situation. If everything else is suitable then an odour may make a small difference but the product of an aerosol spray is unlikely to be the predominant influence and compensate for other failings.'[19]

So is pure French jasmine essence, reportedly the world's most

expensive perfume, really worth up to $300 a gram? More gener-
ally, what effect does a woman wearing any sort of perfume have
on the men around her? The answer seems to depend on how she
is dressed. This at least is what research by Dr Robert Baron of
Purdue University suggests. He told his young male volunteers
they would be helping him investigate the influence of 'first impressions' on character assessment. They were then divided into four
groups. Each group met and chatted briefly with one of two female
assistants. She either did or did not wear perfume and was either
dressed in a blouse, skirt and stockings or informally in jeans and
a T-shirt. When the assistant dressed smartly the perfume made
her seem cold and unromantic. When she wore the perfume with
jeans and a T-shirt, however, the men's impulsive response was to
regard her as warm and romantic.[20] So the response that perfume
evokes depends not just on our sense of smell but also the context
in which it is worn.

## Odours and impulse buys

'Odours act powerfully upon the nervous system,' noted the
17th-century writer Johannes Müller, in *De Febre Amatoria*. 'They
prepare it for all pleasurable sensations, they communicate to it
that slight disturbance or commotion which appears as if insepar-
able from emotions of delight, all which may be accounted for by
their exercising a special action upon those organs whence originated the most rapturous pleasure of which our nature is susceptible.' This holds just as true in the supermarket, department store
or shopping mall as it does in the bedroom, a fact retailers were
quick to realise and exploit in their desire to trigger impulse
purchases. In 1973 the American psychologist Philip Kotler,
Professor of Marketing at Northwestern University, coined the

term 'atmospherics' to describe the 'conscious planning of atmospheres to contribute to the buyer's purchasing propensity' and predicted that 'atmospherics is likely to play a growing role in the unending search of firms for differential advantage.'[21] The discovery that modern consumers enjoy multisensory stimulation while shopping has led to advances in technologies capable of pleasing not only their eyes and ears, but also their senses of smell, taste and touch, by means of lighting effects and plasma screens, specially composed music and aromas to encourage browsing or speed up traffic flow. Today even moderately sized supermarkets allocate space to an in-store bakery despite the fact that it is far more convenient and cost-effective for a central bakery to serve a number of stores. Managers know that, by stimulating hunger pangs, the smell of freshly baked bread encourages people to buy not just bread but also other food, even frozen products.

According to Simon Harrop, chief executive of BRAND sense agency, a British specialist in multisensory marketing, the ability of aromas to create an atmosphere that encourages impulse buying is just as effective when those aromas are created artificially. In the laundry section of a supermarket, for example, shoppers exposed to the scent of freshly laundered sheets not only buy more detergents but also splash out on impulse purchases of products claiming to whiten whites or leave linen smelling like a spring morning. One company has injected the smell of coconut into the shops of a British travel agent, because some suntan oils smell of coconut and the aroma is said to help trigger memories of past holidays and so encourage shoppers to book fresh ones. Overall, research has shown that aromas can make the retail environment more conducive to impulse buying by evoking associations and memories, making the store feel more inviting, stimulating and friendly. They can increase the time people spend browsing the shelves and arouse positive emotions. But our sense of smell is not, of course, our only source of information in the world around us likely to trigger impulsive behaviour.

## The impulsive power of sounds

Sounds may be added to aromas to make the sensory experience even more effective in encouraging impulse purchases. In a supermarket the sale of laundry products significantly increased when a fresh, lemony aroma was combined with the sound of freshly laundered sheets being folded. This, the promoters of the system claim, triggers associations with the comfort, safety and warmth of home. Perhaps by recalling childhood memories of helping mother with the weekly wash.[22]

Music, too, forms an integral part of creating an atmosphere conducive to impulse buying. 'Music communicates with our hearts and minds; it serves as a powerful connection into our emotions,' says Dr M. Morrison of the Department of Marketing at Monash University.[23] 'Music is versatile, it has the ability to relax or invigorate. Music is memorable, it can transport us in an instant to places we want to be. Retailers can use specifically programmed music to create links to past experiences. Music can be a critical component of store atmosphere and plays a role in purchase decision making process . . . retailers can create an audio environment where their customers feel comfortable, relaxed and happy to spend time and money.' In an early study, researchers compared the effects of easy-listening versus Top 40 music on shoppers' estimates of the length of time they had been shopping.[24] Those under 25 reported spending more time shopping when exposed to easy-listening music, while those over 25 thought they had been in the store longer when Top 40 music was played.

Other researchers looked into the most suitable background music to play for selling wine. They began with the assumption that tasting and buying wine is associated with 'higher socio-economic status, prestige, sophistication, and complexity'. They considered that classical music would provide the best 'fit'. In this they were proved right. Impulse purchases of the more expensive wines increased when Mozart, Bach and Vivaldi were played. This, say Richard Yalch and Eric Spangenberg, 'suggests that retailers should devote considerable attention to the symbolic meaning underlying each purchase experience. If consumers are seeking sophistication, then in-store cues must suggest, and even facilitate that experience. The same holds for other sought shopping experiences like excitement, relaxation, etc.'[25] Music preferences, they found, are determined by age rather than the sex of the shopper. 'Middle-aged (25–49) shoppers spent more and shopped longer when foreground music was played, whereas older shoppers (over age 50) shopped longer and purchased more when background music was playing.'

## The impulsive power of warmth

The final sense we will look at is thermoception, or our ability to distinguish between warmth and cold. This can have a very powerful, if seldom appreciated, influence over how we impulsively feel and behave. And it all goes back to our earliest days on earth. Throughout the 400,000 minutes spent in the womb we remain in an environment maintained, provided all goes according to nature, at a constant 37°C. After birth we are kept warm, and fed, by our mother's body. The early experience causes us to form close associations between feelings of warmth, comfort and security. These associations are so powerful that, as adults, they become part of our everyday language. We may say that we have 'warm feelings' for someone or give another the 'cold shoulder'. Unpopular views are said to have received a 'frosty reception', an unemotional person is considered a 'cold fish' and someone we find sexually attractive is 'hot stuff'. We may feel 'lukewarm' about a doubtful proposal and say we have 'gone cold' on one we are about to reject. We may start our first job 'burning with ambition' only to be 'frozen out' by colleagues and so left 'in the cold'. Eventually, the concepts of warmth and social inclusion versus cold and loneliness become so tightly intertwined that we can no longer distinguish between the abstract emotions and the physical sensations. Research has shown, for example, that social inclusion makes us physically warmer just as being excluded by a group leaves us feeling colder.[26] We judge others not only on how emotionally 'warm' they seem but how physically warm we feel.[27] We tend to like people better after meeting them in warm rooms rather than in cold ones. Even holding a warm beverage, however briefly, has been found to increase impulsive liking between strangers.[28] By creating a warm bedroom we help promote passion by increasing feelings of comfort and security. By maintaining just the right temperature inside a supermarket, neither too hot nor too cold, customers can be persuaded to shop for longer and so buy more on impulse. This is a topic I will consider in greater detail in Chapter 10.

All our senses have a role to play in determining when, where, why and how we act on an impulse. Some will trigger a reflex response that bypasses the brain entirely. Accidentally place your hand on a hot stove, for example, and signals from pain receptors will have to travel no further than the spinal cord to cause you to instantly pull it away. Try to lift something too heavy and tension receptors in the muscle will warn you of the fact. Stretch receptors in your gut will alert you when enough has been eaten – although, as we shall see in Chapter 9, not always in time to prevent you from overeating. However, it is those senses described in this and the following chapter whose influence over our impulses is the most profound. They are also, as I have explained, those most likely to be used by others to trigger the impulsive behaviours that they, and not necessarily we, desire.

# CHAPTER 6

# The Power of the Visual

'The brain evolved to predict, and it does so by identifying patterns. When those patterns break down – as
when a hiker stumbles across an easy chair sitting deep in
the woods, as if dropped from the sky – the brain gropes
for something, anything that makes sense. The urge to
find a coherent pattern makes it more likely that the brain
will find one.' Travis Proulx and Steven J. Heine,
'Connections from Kafka: Exposure to Meaning Threats
Improves Implicit Learning of an Artificial Grammar',
*Psychological Science*

In 1978, a woman in New Mexico 'saw' the face of Christ on a
tortilla. Over the weeks that followed, thousands of Catholic
pilgrims trekked from all over the country to pray before it. Scorch
marks had transformed a tasty snack into a holy relic and masti-
cation into veneration. Perceiving patterns, such as a face or an
animal's head, in a stain, a cloud or a piece of food is termed
*pareidolia*.[1] The word – from the Greek *para* (resembling) and *eidos*
(image) – was first introduced into medical practice, around 1885,
by Victor Kandinsky, a Russian psychiatrist.[2] One theory suggests
that pareidolia occurs when individuals who are unable to gain
objective control over their lives try to achieve it perceptually.
'Faced with a lack of control, people will turn to pattern perception, the identification of a coherent and meaningful relationship
among a set of stimuli,' comment Jennifer Whitson and Adam
Galinsky.[3] Such claims illustrate a crucial point about the way we

see the world around us. It is not so much the information coming into our visual senses that matters but what the brain makes of it. Our senses are not just a part of us – they define us. Nothing that we experience throughout our lives would be possible without our senses.

As I showed in Chapter 2, images can trigger an immediate emotional response, ranging from disgust to anger, sadness, horror or outrage. Charities often use emotive images – of starving children or abused animals, for example – in order to trigger an emotional response that will provoke, they hope, an impulse to donate to the cause. One of the first to exploit the power of imagery to stimulate nationalistic impulses was Adolf Hitler. He and his propaganda chief Josef Goebbels exploited every cheap theatrical trick in the book and spared no expense to ensure that impulsive reactions always triumphed over reason. During the Nuremberg rallies, for example, Goebbels set out to manipulate and mesmerise the population through the brilliant use of visual elements. There would be blazing searchlights, towering banners, streamers, flags and standards, hundreds of thousands of men drawn up in highly disciplined formations, goose-stepping marchers, torchlit processions, massive bonfires, human swastika formations and magnificent firework displays.[4] Goebbels also popularised one of the most striking symbols of the modern age in the Olympic torch. Ironically, it was designed to inspire feelings of brotherhood, friendship, the purity of sport and, above all, the athletic superiority of the master race. Every visual art and subterfuge was used to hammer home the message that Nazism was the one true religion and Adolf Hitler its undisputed God.

The power of the visual is demonstrated by a study conducted recently in my own laboratory.[5] Volunteers were asked to complete a survey and 'rewarded' with a piece of delicious fudge. This was so enjoyable that, asked if they would like a second piece, all accepted. This time, however, the fudge had been moulded to look exactly like dog faeces. Despite the fact that it smelled – and would have tasted – as delicious as the fudge they had just enjoyed, not one of the participants would even handle it, let alone put it into their mouth! The 'evidence' provided by their own eyes was sufficiently powerful to generate an impulsive System 1 response that overrode their logical, reflective mind. The participants' disgust was also revealed by their body language, through what are termed 'microgapes'. These facial expressions are so fleeting, typically lasting for less than half a second, that neither the person nor any observers are consciously aware of them. The only way researchers can detect them is via slow-motion video recordings. Despite exerting their influence below the level of conscious awareness, they still strongly affect both participants and onlookers.[6]

More brain power is devoted to processing visual information and, essentially, to creating the world as we perceive it. In the previous chapter I explored smell, sound and thermoception; in this chapter I will explain how what we see, and how we make sense of what we see, is profoundly influenced by culture and experience. Let's start by examining how we see, the mechanics of vision.

## How we see

The basic mechanics of vision, while extremely complex at a molecular level, are now well understood by scientists. Photons, a form of electromagnetic radiation we perceive as light, are focused on our retinas. The retina comprises three layers of cells, one of which contains more than 120 million light-sensitive receptor cells, the rods and cones. Rods and cones contain various proteins ('opsins') that are associated with 11-cis retinal, derived from retinol, or vitamin A. When struck by photons the protein (opsin) changes shape, from kinked to straightened, and this initiates the propagation of a nerve impulse, which is detected by the visual cortex. Membrane proteins act as antennas, reading light signals in this instance, and resonate with the appropriate frequency. The process can be likened to two tuning forks of the same pitch: when one is sounded the other will vibrate as well, due to the resonant energy transferred between the two.

Rods contain rhodopsin, which (because it absorbs green-blue light) appears reddish-purple, hence its alternative name of 'visual purple'. Rhodopsin is responsible for *monochromatic* night vision. The cones have opsins sensitive to slightly different wavelengths of light. Photopsin III, with a peak at 430 nanometres, produces the colour we subjectively interpret as blue-violet. Photopsin II, with a peak at 530 nanometres, is sensitive to blue-green while Photopsin I, with a peak at 560 nanometres, is perceived as yellowish-green. This means, of course, that colour exists *only* in the brain and not in the world around us. The green of a leaf or the red of a tulip are created by what happens behind our eyes rather than in front of them.

The retinas convert light energy into electrical signals which travel as impulses, via the optic nerve, to two peanut-sized clusters of cells, the lateral geniculate bodies, deep inside the brain. From here the signals travel to the striate or primary visual cortex, at the rear of the brain, and are then sent onwards to several other higher visual regions. Neuroscience currently, and most likely for decades to come, can provide no answer as to how neural impulses are transformed into *qualia,* our subjective responses to the glories of a brilliant orange and red sunset or the golden beauty of a 'host of golden daffodils'. But while sighted people are born with the ability to see at the moment their eyes are first opened they cannot perceive the visual world as they will come to perceive it.

## We have to learn how to see

After Sidney Bradford got his sight back at the age of 52, he was only able to make sense of the world through his finely tuned sense of touch. 'Bradford's responses . . . were far from normal,' reports pioneer vision researcher Richard Gregory. 'Pictures looked flat and meaningless. Perspective meant nothing to his visual system, yet he could judge the distances and sizes of objects that were already familiar from touch, such as chairs scattered around the ward.'[7] The difficulties facing someone whose sight has suddenly been restored is further illustrated by the case of S.K. He was born in Bihar, India with a rare genetic condition in which the lenses are almost completely absorbed into the interior of the eyes. Poverty prevented surgery while he was an adolescent and he was educated in a school for the blind. At the age of 29 his world of darkness came to an end when corrective optics were implanted. For many weeks after the operation he had great difficulty recognising even simple shapes, such as triangles and squares. 'His brain was unable to distinguish the outlines of a whole shape; instead, he believed that each fragment of a shape was its own whole,' the researchers reported. 'When three-dimensional shapes, such as cubes or pyramids, were shown with surfaces of different luminance consistent with lighting and shadows . . . [he] reported perceiving multiple objects, one corresponding to each facet . . . it seems like the world has been broken into many different pieces.'[8] As these cases demonstrate, seeing – as opposed to being sighted – is a skill that has to be learned. A person blind from birth whose sight is restored by surgery must slowly and painstakingly master the rules by which to make sense of the torrents of incoming visual information.

Growing up in a particular culture we unconsciously absorb the visual 'rules' by which to interpret the signals from our retinas.

Within a few weeks of birth this process has become so automatic and effortless that we take everything we see, or think we see, for granted. Although the image on the retina is two-dimensional, we perceive a three-dimensional world mainly, but not solely, through a process known as *stereopsis*. This process, first described by Sir Charles Wheatstone in 1838, depends on each eye receiving slightly different images and the brain then fusing these separate images into one to create a vivid impression of depth. Other rules create the impression of depth and allow us to judge the size of everyday objects. If something we know to be fairly large (such as a house, for example) appears smaller than expected, a rule tells us this is because the house is seen at a distance rather than being a miniature building.

As with many visual rules this one is easily fooled, as illustrated by the Ames Room illusion shown in figure 1.

*Figure 1: The Ames Room illusion*

This appears to show a giant and a midget. In fact, despite the evidence of our own eyes to the contrary, both are the same height. The room has been constructed so that it is taller and deeper on the left, making the leftmost child look short and the man on the right seem huge because he is standing on the smaller side of the irregularly shaped room. However, because our normal experience of rooms tells us that all the walls are the same height whereas people differ very greatly in height, it is this rule that the brain follows.

Parallel lines, such as railway tracks, appear to come closer and closer together as they run off into the distance, an example of perspective, one of the most powerful indicators of depth.

*Figure 2: The Ponzo illusion*

This forms the basis of the Ponzo illusion (figure 2), where the two lines appear to be receding into the distance and the horizontal line at the bottom of the picture appears shorter than the line at the top, although both are the same length.

The Kaniza triangle illusion shown in figure 3 is a typical example of what happens when the brain seeks to apply rules that just don't work. How many triangles are depicted here? Two, six, eight or none?

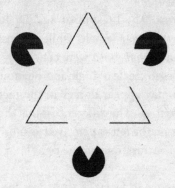

*Figure 3: The Kaniza triangle illusion*

The surprising answer is that there are no triangles. Just three V shapes and three shapes that look a little like Pac-Men. Your brain, blindly following visual rules, has merely invented them.

Contrast rules can also lead to errors. A gargoyle jutting out from the side of a building will be lighter if lit from above. Since this is how most things are lit, another low-level processing rule says that this is how it must always be, even when it isn't. Judgements based on the contrast rule can lead to some curious perceptions, as the images in figure 4, created by Dr Richard Russell from the Department of Psychology, Harvard University, demonstrate.[9]

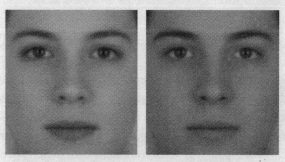

*Figure 4: 'The Illusion of Sex'*

In what he terms 'The Illusion of Sex' Dr Russell has made the face on the left appear female, while the image on the right appears male even though both were created by morphing male and female features to create a single androgynous face. Although the eyes and lips may appear darker in the picture on the left, in fact both are identical. The difference is that the higher contrast and lighter image on the left is perceived as more feminine while the darker, lower contrast, one on the right has a more masculine look.

'This sex difference in facial contrast was found in both East Asian and Caucasian faces,' comments Richard Russell. 'Female or male faces with greater facial contrast were rated as more feminine or less masculine than faces with less contrast, though the relation was very weak for male faces. Decreasing or increasing the facial contrast in an androgynous face is sufficient to make it appear male or female. These findings indicate that, while people are not consciously aware of this sex difference, their perceptual systems nevertheless make use of it. Because femininity and attractiveness are strongly related these results help explain an earlier finding that female faces are more attractive with increased facial contrast than with decreased facial contrast.'[10]

## The Power of Subliminal Priming

While the overt potency of certain symbols, such as national flags, the Red Cross and Red Crescent, the crucifix and even McDonald's Golden Arches, has long been recognised and exploited, the impact of briefly displayed images has only recently been discovered. In a process called 'priming' volunteers are shown pictures for a very short time, just a few thousandths of a second, so that they are

perceived subliminally, or below the threshold of conscious awareness. Despite its very rapid exposure the image still has the power
to influence behaviour through System I thinking, as the following
two studies demonstrate.

In research aimed at exploring the extent to which analytical
(System R) thinking promotes religious disbelief, Will Gervais and
Ara Norenzayan from the University of British Columbia showed
their undergraduate volunteers either two images illustrating deep
thought – such as Rodin's statue *The Thinker* (figure 5) – or two
pieces of control artwork. These were pictures, like the Greek
sculptor Myron's famous discus thrower (figure 6), matched for
external characteristics, such as colour and posture.[11]

*Figure 5: Rodin's* The Thinker; *Figure 6: Myron's famous discus
thrower*

Remarkably they found that even a brief glimpse of images
promoting System R thinking enhanced rational thought, despite
the fact that none of their volunteers were aware of how they had
been influenced.

A very different but equally revealing study was carried out by
Chen-Bo Zhong and Sanford DeVoe at the University of Toronto's
Rotman School of Management. They showed fast-food logos,

from McDonald's, KFC, Subway, Taco Bell, Burger King and Wendy's, to their student participants for just 12 thousandths of a second. Although none of the subjects were consciously aware of what they had seen, they became less patient and showed an increased preference for time-saving products, such as 2-in-1 shampoo, over regular products. 'Fast food represents a culture of time efficiency and instant gratification,' comments Chen-Bo Zhong. 'The problem is that the goal of saving time gets activated upon exposure to fast food regardless of whether time is a relevant factor . . . we're finding that the mere exposure to fast food is promoting a general sense of haste and impatience regardless of the context.'[12]

## How culture influences what we see

The way we make sense of the world is strongly influenced by rules imposed by the culture in which we grew up. 'Each individual's experiences combine in a complex fashion to determine his reaction to a given stimulus situation', says Marshall Segall, Professor Emeritus of Social and Political Psychology in the Maxwell School at Syracuse University. 'To the extent that certain classes of experiences are more likely to occur in some cultures than in others, differences in behaviour across cultures, including differences in perceptual tendencies, can be great enough even to surpass the ever present individual differences within cultural groupings.'[13] In a study exploring these differences, Takahiko Masuda of the Department of Psychology, University of Alberta and Professor Richard Nisbett, from the University of Michigan, asked Japanese and American subjects to watch a total of ten 20-second animations depicting underwater scenes with fish of various sizes, aquatic plants and rocks.[14] Having viewed the animations the participants then answered questions, from memory, on what they had seen. The researchers found that while the Americans had directed their gaze primarily towards the brightest or most rapidly moving fish, the Japanese were more likely to report seeing the stream, noting that the water was green and there were rocks on the bottom, before ever mentioning the fish. Overall they provided 65 per cent more information about the background than the Americans did and twice as much about the relationship between background and foreground objects.

Another study used an eye-tracker to identify differences in the way European Americans and native Chinese viewed pictures. Eye-trackers use beams of infrared light to detect precisely where a person is looking, the direction their gaze takes as they view the image and how long it remains fixated on any one area. Two groups of graduate students, half of them Chinese and half Americans of European descent, took part in the study. Their eye movements were closely monitored as they looked at various pictures, two of which are shown in figures 7 and 8.

*Figure 7: A train in a railway yard*

*Figure 8: A bird by a bridge*
*The train and the bird are considered the focal objects.*
*(From Boland, Chua and Nisbett (2005) with permission)*

As in the earlier study, the Americans looked at focal objects (i.e. the train in the railway yard and the bird by the bridge) sooner and gazed at them for longer than the Chinese. The latter made a greater number of eye movements, shifting their attention back and forth between the main object and the background. Reinforcing the notion that such differences are cultural, a study by Julie Boland, Hannah Chua and Rick Nisbett at the University of Michigan identified distinct differences in the way American and Chinese students moved their eyes when viewing the same scenes. 'In sum, we did indeed observe cultural differences in the fixation patterns of Chinese and American participants,' the researchers reported, 'suggesting that we literally see the world differently . . . to be clear, we are not suggesting that there are cross-cultural differences in object recognition (i.e., recognising the central object as a cow) or colour perception, or the way in which visual information travels through the visual pathways in the brain. We assume that all of those basic vision processes are universal. The point is that the same stimulus is attended differently by people from different cultures.'[15] One explanation is that Asian communities are more socially complex than those of Americans. It is argued that the key feature of Chinese and Japanese cultures is a desire for consensus whereas Western culture is more concerned with getting things done and getting ahead, even at the expense of social harmony. The roots of these differences may lie in the distant past. Because rice farmers in ancient China depended on complicated, communally shared irrigation systems they had to get along with one another if they were to survive. By contrast Western culture first developed in ancient Greece, where agriculture was individual rather than communal, with people running their own farms, growing grapes and olives, and functioning as businessmen. 'Aristotle focused on objects. A rock sank in water because it had the property of gravity, wood floated because it had the property of floating,' comments Professor Nisbett. 'He would not have mentioned the water. The Chinese, though, considered all actions related to the medium in which they occurred, so they understood tides and magnetism long before the West did.'[16]

What makes these findings especially interesting is that they demonstrate differences in the low-level perceptual processes by which we control our eyes. How we make sense of what we see strongly influences our behaviour, frequently causing us to behave impulsively rather than rationally. But although powerful, our visual sense is by no means the only one that triggers the impulse. These studies illustrate the extent to which our understanding of the visual world is strongly dependent on learning. While culture clearly exerts a major influence over this process, so too do aspects of the mental tasks we undertake.

## How making sense affects our visual sense

Czech-born Franz Kafka, author of *The Metamorphosis*, *The Trial* and *The Castle*, gave the world the term Kafkaesque, meaning 'characterised by senseless, disorienting, often menacing complexity'. According to Albert Camus the fundamental ambiguity of Kafka's writing arises from its 'perpetual oscillations between the natural and the extraordinary, the individual and the universal, the tragic and the everyday, the absurd and the logical'.[17] Drs Travis Proulx of the University of California at Santa Barbara and Steven Heine, from the University of British Columbia[18], decided to make use of this ambiguity in order to see whether coping with it was causing people to perceive the world differently. They asked a group of students to read Kafka's short story *The Country Doctor* before presenting them with a series of six to nine long strings of letters, for example, XMXRTVTM, VTTTTVM. The students were then shown 60 further letter strings and asked to identify any they had seen before. The two sets of letters were related, in a very subtle way, with some more likely to appear before or after the others. The same task was also given to a second group of students, matched for age and ability, who had read not Kafka but a story of the same length, written by Proulx and Heine, that contained no ambiguities. The results were intriguing. Those who had read Kafka's short story identified around 30 per cent *more* letter strings and were almost twice as accurate in their choices than the group who had read a less ambiguous piece of prose. 'Two general conclusions can be drawn from these findings,' comment the authors. 'First, the breakdown of expected associations presented in the absurd story appeared to motivate participants to seek out patterns of association in a novel environment. Second, and more remarkably, participants in the meaning-threat condition demonstrated greater accuracy in identifying the genuinely pattern-congruent letter strings among the test strings, which suggests that the cognitive mechanisms responsible for implicitly learning statistical regularities in a novel environment are enhanced by the presence of a meaning threat.' The way we make sense of new information is, therefore, strongly influenced by personal biases.

Research has shown that, after being asked to reflect on their own mortality, people become more religious, more patriotic and less tolerant of strangers. If insulted they express greater loyalty to friends and if informed they have done poorly on a quiz they identify more strongly with the winning team. This suggests that built-in rules are used to try and make sense of information that can be interpreted in more than one way. System R follows stringent rules, similar to those of traditional logic. System I, by contrast, uses associations, either inborn or learned, which make it faster but less precise. 'Attention is the key to distinguish between unconscious thought and conscious thought,' say Ap Dijksterhuis and Loran Nordgren from the University of Amsterdam, who have pioneered this fruitful area of research. 'Conscious thought is thought with attention; unconscious thought is thought without attention (or with attention directed elsewhere). However, this does not mean that conscious thought comprises only conscious processes. One could compare it to speech. Speech is conscious, but various unconscious processes (such as those responsible for choice of words or syntax) have to be active in order for one to speak. Likewise, conscious thought cannot take place without unconscious processes being active at the same time.'[19] In the next chapter we will look in more detail at some of the key personality variables that influence our perceptions of the world and the ways in which these encourage impulsive risk taking.

# CHAPTER 7

# Impulses and the Risk-Taking Personality

'If financial decisions and capital markets are driven by rational thinking, then why does the manner in which the crisis developed seem so irrational?' Andrea Rinaldi, *Homo economicus*, EMBO Reports

During the second week of September 2008, while conducting research in New York, I stayed in a hotel next door to the Lehman Brothers building. From my bedroom window I could look down on a flat roof, covered in Astroturf, used by their staff for exercise. For several days I watched the same man chasing a ball up and down the length of the roof, running endlessly backwards and forwards through the early morning light. At the time I wondered at his energy, the monotony of his routine and his skill in controlling the ball. I might have done better speculating about the lengths of his fingers! Knowledge of his digit lengths could have alerted me to events of 15 September when the 158-year-old investment bank went spectacularly bust. With debts of $619 billion it was the biggest bankruptcy in history and the subsequent shock waves accelerated the collapse of the entire banking system. 'As with every complex phenomenon, a mixture of causes have been put forward,' comments Andrea Rinaldi of the European Molecular Biology Organisation, 'the housing-market bubble, risky mortgage loans, "toxic" financial products, high personal and corporate debts, inaccurate credit ratings, and much more.' That 'much more' includes a culture of risk taking among the traders and bankers themselves. A culture which, while no doubt turbocharged by greed, has biology as its main driving force. Innate factors endowed those responsible with genetic and endocrinal reasons for their incredible and reckless appetite for risk.

When John Coates and Joe Herbert, from the University of Cambridge, measured levels of endogenous testosterone and cortisol in a group of male, City of London traders, they found that testosterone was higher than average on days the trader made more than usual. Levels of the stress hormone cortisol, by contrast, rose on days he experienced increased uncertainty of return, investment risk and market volatility. 'Testosterone and cortisol are known to have cognitive and behavioural effects,' say the researchers, 'so if the acutely elevated steroids we observed were to persist or increase as volatility rises, they may shift risk preferences and even affect a trader's ability to engage in rational choice.'[1]

In a more recent study, John Coates and colleagues measured the lengths of male 'high-frequency' traders' second and fourth digits.[2] (High-frequency or 'noise' trading involves monitoring the computer screens continuously and responding almost instantly to oscillations in prices.) They found that the traders' appetite for risk taking could be accurately predicted by comparing the lengths of these two fingers. Which is where my speculations about the solitary ball-player on the roof of the Lehman building come in. Research has shown that men who are impulsive risk takers have a ratio of less than 1 if the length of their index finger (second digit) is divided by their ring finger (fourth digit).[3] This has been dubbed the 2D:4D ratio. The ratio differs reliably by sex, with males typically having a lower 2D:4D on their right hand than on their left hand. This difference is found in children as young as two and may be established in the womb by the 13th or 14th week after conception.[4] It is known that prenatal androgens affect brain development by increasing its sensitivity to testosterone later in life. In this way they shape such behavioural traits as a preference for risk, heightened vigilance and faster reaction times. Several markers have been suggested for assessing levels of prenatal androgens but of these the 2D:4D ratio has proved one of the most reliable. A relatively longer fourth finger indicates higher prenatal androgen exposure. The 2D:4D hypothesis of Coates and his colleagues was that a higher prenatal testosterone exposure would improve a trader's performance: traders with a lower 2D:4D ratio would make greater long-term profits and would remain in the business for a longer period of time. Intriguingly, their predictions were fulfilled. The results also showed that the lower a trader's 2D:4D ratio, the more money he made as the market increased. '[I]f markets select traders on the basis of their profitability and their occupational preferences, then low-2D:4D traders will continue to influence asset prices and equilibria in some of these sectors. Contrary to the assumptions of the rational expectations hypothesis, financial market equilibria may be influenced as much by traders' biological traits as by the truth of their beliefs.' the authors remarked.[5] Due to its early appearance, the most likely explanation for the 2D:4D ratio is that prenatal exposure to the male hormone testosterone not only slows the growth of the index finger relative to the other three fingers, excluding the thumb, but also influences the development of the fetal brain. As a result, finger length offers a (literally) handy measure of prenatal testosterone exposure. While still somewhat speculative, this theory is supported by two findings. The first is that the same genes (named Hoxa and Hoxd) underlie the development of both fingers and testicles.[6] Hox genes determine an organism's basic structure and orientation. The second is that genetically related variations in sensitivity to androgen influences digit ratio. The 2D:4D ratio has been found to correlate with several psychological traits in addition to impulsivity, risk taking and aggression.[7]

The crucial role played by genes in determining someone's level of risk taking has also been studied by Jonathan Roiser and his colleagues at University College London. They investigated the influence on economic decision-making by genes involved in activating the amygdala, a region of the brain involved in processing emotions. 'Our research demonstrates that an individual's genetic make-up influences an economic decision-making bias known as the "frame effect",' explains Roiser. 'The frame effect occurs when the phrasing (or "framing") of a decision affects an individual's eventual choice, even when the meaning of the decision is not changed. So, for example, a supermarket might advertise their yoghurt as "99 per cent fat-free" as opposed to "1 per cent fat", though these two statements mean the same thing.'[8]

In one of their studies, volunteers were asked to decide whether or not to gamble £50 with a 'gain frame' and a 'loss frame'. The 'gain frame' offered two options. Option A was to keep £20 and option B to gamble, with a 40 per cent chance of keeping the full £50 and a 60 per cent chance of losing everything. In the 'loss frame', they could elect to lose £30 for certain (Option A) or take the same gamble as in the 'gain frame'. Individuals with a specific genotype were significantly more susceptible to the framing effect than were those possessing a different genotype. 'In this study, we were able to demonstrate that the amygdala was more active during decisions where the frame influenced an individual's choice in carriers of the genotype more vulnerable to the bias,' explains Roiser. 'This suggests that the bias . . . may have been driven by automatic emotional responses to the framing of the question, over-riding analytic decision-making processes', which occur in other brain regions. Away from the laboratory, this would mean for example that traders, required to assess risk rapidly and ensure their decisions remain consistent irrespective of the way information is presented to them, would perform better if they had one genotype rather than another.

Women with smaller digit ratios tend to display more masculine interests and traits. Female observers report that men with smaller 2D:4D ratios are more masculine and dominant.[9] Researchers have also reported significant correlations between more masculine (small) digit ratios and achievement, ability and speed in a variety of sports and in visual-spatial ability. However, male-like digit ratios are also associated with increased rates of autism, immune deficiency and reduced verbal fluency.[10] In a study led by Dr Gad Saad of Concordia University and the John Molson School of Business, the index fingers of some 400 male and female students were compared with their ring fingers. Researchers reported that those men, but not women, with low ratios were more likely to act impulsively and engage in risky activities. Dr Saad commented: 'Our findings show an association between high testosterone and risk-taking among males in three domains: recreational, social and financial.'[11]

Why are the effects found only in men? The researchers suggest a possible explanation lies in the fact that, unlike males, women do not use risky behaviour as a mating signal. According to Camelia Kuhnen and Joan Chiao of Northwestern University, Illinois, impulses that now play a crucial role in financial decision-making may have arisen from 'evolutionarily adaptive mechanisms that encourage novelty-seeking behaviour', for example during migrations to new territories, while exploring unfamiliar terrain, when hunting or when competing for sexual partners.[12] Should this evolutionary perspective turn out to be correct, then the 'animal spirits' that some economists have suggested are the real key to understanding how and why an economy fluctuates might, in part, be deeply rooted in the relationship between our genes, our hormones and the neurological pathways neuroscience is starting to uncover.

## Discover your own risk ratio

To calculate your 2D:4D ratio all you have to do is to measure the second and fourth digits of your dominant hand (i.e. right hand if right-handed) as shown in figure 1. Once you have done that divide the length of your second digit by that of your fourth. The table below will provide you with your likely attitude towards risk taking.

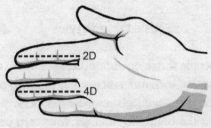

*Figure 1: Illustration of 2D:4D ratio measurements*

Here are two examples that will help to make the calculation clear:

Example 1
Length of second digit = 6.1 cm
Length of fourth digit = 7.3 cm
Ratio 2D/4D = (6.1/7.3) = 0. 83
Risk-taking impulse: high

Example 2
Length of second digit = 7 cm
Length of fourth digit = 6.5 cm
Ratio 2D/4D = (7/6.5) = 1.1
Risk-taking impulse: fairly low

| Ratio of second to fourth digit | Impulsivity level |
| --- | --- |
| Less than 0.83 | High |
| 0.84–1.0 | Moderate |
| 1.1–1.02 | Low |
| Greater than 1.02 | Very low |

## Risk-taking styles

While some people thrive on risk, others shun it. As we shall see later in this chapter, such differences are most likely to result from the levels of certain chemicals, namely oxytocin, serotonin and testosterone, in the brain. To discover your own preferred risk-taking style, answer the seven questions below. Although people will sometimes change their style of decision-making according to specific circumstances, some of which I discuss below, we generally feel most comfortable with one particular style.

**1.** It is early morning and you are about to leave for work. Your job is going to take you outdoors a good deal today and, although the sky is clear at the moment, the forecast has warned of rain. Because you have other things to carry, it will be awkward to take an umbrella along and you do not want to wear a raincoat unnecessarily. Which of the following thoughts is most likely to be uppermost in your mind as you prepare to leave the house?

**(a)** Even though the forecast was bad, the sky looks clear to me and that's usually the sign of a fine day. I'll take a chance and leave my umbrella and raincoat at home.
**(b)** It always seems to rain when I leave my umbrella and raincoat

at home, so, rather than risk a soaking, I'm going to have to take them along despite the inconvenience.

(c)  I would feel foolish if I were caught in a downpour without any protection. On the other hand, I shall be annoyed if I burden myself unnecessarily and it doesn't rain. I'll compromise by taking my umbrella and hope this will keep me sufficiently dry if there is a storm.

2.  You decide to invest on the stock market and your broker mentions three different companies in which you might be interested. One offers a safe investment but little chance that it will ever bring in a large profit. The second is a mining company whose shares some pundits expect to show a spectacular rise in price but which is highly speculative. Finally you could buy shares in a small manufacturing company whose board of directors are considering a takeover by a multinational. If this happens, the stock will rise in price; if not, the price is going to remain where it is. When deciding which stock to purchase, which of these thoughts is most likely to be passing through your mind?

(a)  It's a risk, but I'll buy mining stock. Who dares wins!

(b)  With my luck, if there is any possibility of the stock going down, it will. I'll opt for the safe blue-chip investment and a small but steady return.

(c)  I'd hate to miss out on a good deal. On the other hand, I'll kick myself if I lose money or only break even when there is profit to be made. I'll go for the manufacturing company because, even if the worst happens, I won't feel too bad about my choice.

3.  You have been moved by your firm to another part of the country and obliged to place your home on the market for £480,000. Because you have had to borrow money at a high interest rate to purchase a

new house, it is important to sell your old one as quickly as possible. Although your asking price is a fair one, the property market is suffering a downturn and houses in your area tend to sell slowly. The day after your house goes on the market a prospective purchaser arrives and offers £430,000 in cash. By accepting £50,000 less than your asking price you will clear your debt on the new home but still feel hard done by. However, if you hang on for the asking price and the house fails to sell within the next three months, you will pay more than that in interest charges. As you reflect on the offer, which of the following thoughts is likely to be running through your mind?

(a) I would be crazy to accept £50,000 less than my fair asking price. No deal.

(b) I'll grit my teeth and accept the offer. With my luck and the bad state of the market, it may be a long time before I get anywhere near as good an offer again.

(c) If I accept this offer, I shall regret it; on the other hand, if I turn it down flat and nothing better comes along for half a year, I shall feel bad about that decision, too. I will try to keep this buyer interested for a few weeks by stalling and make every effort to sell the house at the price I want in the meantime. If I don't appear too keen, he may up his offer a bit, and, at the worst, I shall have a buyer to fall back on if I can't get the price I want.

**4.** Your firm's prosperity depends to a great extent on the drive, enthusiasm, and success of its sales force. A rival organisation has fired their top salesman and you know that his experience, skill and contacts could prove a tremendous asset. The difficulty is that you cannot find out for certain why he was dismissed. His version is that there was a personality clash between himself and the new sales director, but the grapevine suggests there might have been other reasons. One story is that he has a drinking problem and is no longer trustworthy; another claims he was caught defrauding the company and only just avoided prosecution. After interviewing him you still do not know the truth, but must decide whether or not to offer him employment. While pondering your decision, would you be most likely to reflect that:

(a) It would be worth taking a chance because he could be tremendously valuable, provided the rumours are untrue. The malicious gossip could have been spread by the sales director trying to get back at an awkward ex-employee and one should always trust one's own judgement when assessing staff.

(b) It would be foolish to take a risk because where there's smoke there is almost always fire, and even if those stories are untrue, the real reason for his dismissal could be equally serious. It is better to be safe than sorry and accept that if things can go wrong they probably will go wrong.

(c) It would be best not to rush into a decision which is likely to give me cause for regret later on. If I employ someone who is dishonest or unreliable, the mistake could be serious; on the other hand, it would be foolish to turn his application down flat because, if he is reliable and honest, I don't want to lose him to a rival. My best course of action is to tell him I cannot come to an immediate decision but will let him know by the end of the week. In the meantime I can take steps to check out the allegations more carefully and see if there is any truth in them.

5. After leaving secondary school, your daughter was faced with a difficult career choice. She was offered a place at a prestigious university where getting a good degree would prove hard but would be a considerable asset on the job market should she succeed. Alternatively, she could go to a less distinguished university where a good degree might be easier to obtain but lack the same

distinction. In the event she chose to go to the prestigious university but ended up with a third-class degree. Do you think:

**(a)** She made the right choice by aiming high because you should always go for the best there is. Her failure is unfortunate, but it need not prevent her from doing well in the real world.

**(b)** She only has herself to blame for what happened because over-reaching oneself inevitably results in failure and disappointment.

**(c)** Her best approach would have been to apply to a wide enough range of universities to avoid a disappointing outcome. By doing so she might have found one that combined a fair amount of prestige with a course better suited to her intellectual abilities.

**6.** You are owed £2,000 and feel your chances of recovering the money are slender, since the debtor seems to have no assets and few prospects of ever obtaining the sum. When you confront him, he frankly admits that he is almost broke but offers you three choices. He will give you all the cash he possesses, some £1,200, in full and final settlement of the debt. Alternatively, being a gambling man, he is prepared to offer a wager based on the toss of a coin. If he wins, you must forget about the debt and give him a letter to that effect. If you win, however, he will part with the only thing of any value he possesses, a solid gold watch which belonged to his father and has been valued by a reputable jeweller at £5,000. Are you most likely to think:

**(a)** I'll take a chance on the coin toss. After all, I am only really risking £1,200 since I am never likely to get more out of him than that. If I win the watch I could more than double my original investment.

**(b)** Knowing my rotten luck I would be bound to lose. It's crazy to risk everything on the toss of a coin. I'd better write off part of the debt, accept his offer of £1,200 and learn not to be so trusting when lending money in the future.

**(c)** I cannot agree to any of his offers because, whatever I decide,

I shall end up feeling bad about it. If I win, I will feel I have taken advantage of someone in trouble and profited unreasonably from his misfortune. If I lose, I shall be mad at myself for throwing away all that money. If I accept his settlement, I shall always believe I should have done better. So I'll tell him there's no deal and he must come up with some way of paying me the full amount.

**7.** You have been given tips on three horses running in different races. As the information comes from a good friend who is also an experienced trainer, you decide to bet £100 on the first horse, which romps home at odds of ten to one. With the remaining horses yet to run, would you be most likely to decide:

**(a)** I'll put my money on the second horse and, if that comes in first, transfer those winnings to the third. I would be crazy to pass up this chance to make a really big win.

**(b)** I'll collect my money and go home while I'm still ahead. Nobody knows enough about horses to predict three winners in a row and, with my luck, I'll lose the whole lot.

**(c)** I'll put some of my winnings on the next two tips. I would be annoyed if I failed to bet and they were first past the post. On the other hand, it would be foolish to risk all my gains on the next two races.

## How to score

Total the 'a's, 'b's and 'c's, and note which response appears most frequently. If you have a majority of 'a's, then your risk taking-style is that of a Maximiser; a majority of 'b's indicates you are a Minimiser; a majority of 'c's indicates you are a Mini-Maxer. If two or more scores are close, then your style is likely to vary from situation to situation. Each of the three styles has certain strengths and weaknesses. By understanding what motivates you to respond in a particular way, you can ensure that your approach matches the particular situation. It also enables you to identify the tactics being used by others and so take advantage of their weaknesses or appreciate their strengths.

**Maximiser**: This favours outcomes promising the greatest gains should things work out as planned. Maximisers are perpetual optimists with an unshakeable belief in their ability to eventually emerge triumphant, no matter how seemingly risky the situation. Little time is wasted on regrets or recriminations when a wrong decision is made. It is the risk-taking profile typical of impulse-driven entrepreneurs, always ready to stake their all on the chance of winning big.

**Minimiser:** Highly risk averse, Minimisers seek to reduce the possibility of incurring any loss to a minimum. Like the famous Captain Murphy, who originated Murphy's Law, they firmly believe that 'if something can go wrong it will go wrong' and therefore they must adopt the course of action that best protects them from loss or helps minimise any losses should the worst happen. Minimisers are reluctant to commit themselves until all the risks have been fully identified and analysed. This makes them especially vulnerable to 'the paralysis of analysis', a state of procrastination in which nothing ever seems to get done. It is a style best suited to situations in which there is a strong possibility of material or financial loss.

Mini-Maxer: The aim here is to seek out the option that offers to minimise any probable loss while maximising any possible gain. In other words, to choose the option which ensures the best possible results should things go according to plan but safeguards against too many regrets if events fail to match up to expectations. Those using this style are highly sensitive to differences between what was accomplished and what might have been achieved. They regard missed opportunities as a source of regret and frustration, so always seek to keep them to a minimum.

As I have explained, we tend to adhere to our favoured approach to risk taking in all aspects of our lives. Under certain circumstances, however, we may switch from low or moderate to high risk taking. It depends on how trusting we feel and this, in turn, is directly related to our brain chemistry. There is, of course, a direct link between the levels of risk we feel comfortable exposing ourselves to and our tendency to be impulsive. In general Maximisers tend to behave far more impulsively that Minimisers, with those who adopt a Mini-Max approach falling somewhere between the two.

## How impulsive are you?

Here are two different types of questionnaires which will enable you to gain insights into the extent to which you act on impulse.

The first, created by my colleague Dr Margaret Yufera-Leitch, involves matching symbols and will give you a general indication of whether or not you are impulsive. This is a shortened and simplified version of a computer-driven test that you can find online at: http://www.impulsive-eating.com/measuring-impulsivity.html

The second questionnaire was created by myself and has been, at the time of writing, completed by around 10,000 people. This offers a more graded view of impulsivity, rather than the binary either/or result provided by the first test.

### Questionnaire 1
Your challenge in this self-test is to identify which of the four symbols on the *right* matches the one on the left. Time yourself and then check your answers against the score chart on page 104.

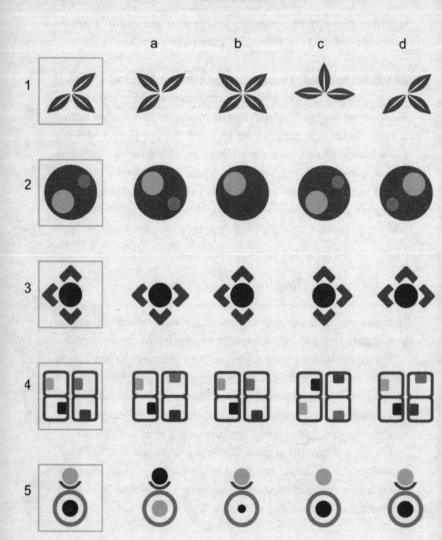

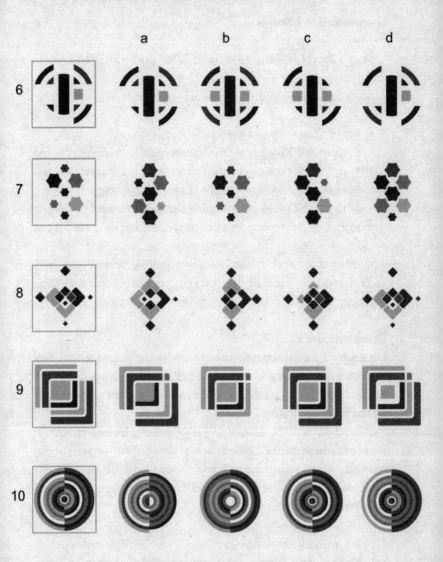

## Questionnaire 1 Scores

Correct answers:

(**1**)d  (**2**)c  (**3**)b  (**4**)b  (**5**)d  (**6**)a  (**7**)b  (**8**)d  (**9**)c  (**10**)c

Score: You are impulsive if you achieved:

- Fewer than 6 correct choices
- 6  right AND took *more* than 12 seconds to complete the test
- 7  right AND took *more* than 14 seconds to complete the test
- 8  right AND took *more* than 16 seconds to complete the test
- 9  right AND took *more* than 18 seconds to complete the test
- 10 right AND took *more* than 20 seconds to complete the test

Any other score and you are *not* an impulsive person! You took your time and studied the target stimulus accurately and correctly.

## Questionnaire 2

For each of the 20 statements below, note the response that best describes the way you would act or feel in the situation described. Answer rapidly and truthfully, since your immediate response will be the most accurate.

(This test, like the first, is available online, free of charge, at www.impulsive.me.uk. When using the online version, your answers, your accuracy, and your speed will be dynamically analysed and the score obtained will, as a consequence, be more scientifically accurate.)

1. I act without thinking.
(**a**) Always  (**b**) Often  (**c**) Occasionally  (**d**) Rarely or never

**2.** I blurt things out.

(**a**) Always  (**b**) Often  (**c**) Occasionally (**d**) Rarely or never

**3.** I quickly make up my mind.

(**a**) Always  (**b**) Often  (**c**) Occasionally (**d**) Rarely or never

**4.** I am happy-go-lucky.

(**a**) Always  (**b**) Often  (**c**) Occasionally (**d**) Rarely or never

**5.** I have difficulty focusing my attention for long.

(**a**) Always  (**b**) Often  (**c**) Occasionally (**d**) Rarely or never

**6.** I find thoughts constantly racing through my mind.

(**a**) Always  (**b**) Often  (**c**) Occasionally (**d**) Rarely or never

**7.** I would sooner do things on the spur of the moment than plan in advance.

(**a**) Always  (**b**) Often  (**c**) Occasionally (**d**) Rarely or never

**8.** I lack self-control.

(**a**) Entirely  (**b**) To some extent (**c**) Not to any great extent

(**d**) Not at all

**9.** I take life as it comes.

(**a**) Always  (**b**) Often  (**c**) Occasionally (**d**) Rarely or never

**10.** I get fidgety watching a lengthy film or play.

(**a**) Always  (**b**) Often  (**c**) Occasionally (**d**) Rarely or never

**11.** When given a problem I rush to find an answer.

(**a**) Always  (**b**) Often  (**c**) Occasionally (**d**) Rarely or never

**12.** I leave my financial future to take care of itself.

(**a**) Always  (**b**) Often  (**c**) Occasionally (**d**) Rarely or never

**13.** I dislike tackling complex problems.

(**a**) Always  (**b**) Often  (**c**) Occasionally (**d**) Rarely or never

**14.** I change jobs.

**(a)** Constantly **(b)** Frequently **(c)** Occasionally **(d)** Rarely or never

**15.** I am easily bored.

**(a)** Always **(b)** Often **(c)** Occasionally **(d)** Rarely or never

**16.** I am impatient.

**(a)** Always **(b)** Often **(c)** Occasionally **(d)** Rarely or never

**17.** I act after careful deliberation.

**(a)** Rarely or never **(b)** Occasionally **(c)** Usually **(d)** Always

**18.** I buy things on impulse.

**(a)** All the time **(b)** Frequently **(c)** Occasionally
**(d)** Rarely or never

**19.** I have difficulty multitasking.

**(a)** Always **(b)** Often **(c)** Occasionally **(d)** Rarely or never

**20.** I quickly lose interest in new hobbies or pastimes.

**(a)** Always **(b)** Often **(c)** Occasionally **(d)** Rarely or never

Score each response as follows: **(a)** 4 points; **(b)** 3 points; **(c)** 2 points; **(d)** 1 point.

| Total score | General comments |
| --- | --- |
| 20–25 | You are far less impulsive than most people and like to reflect coolly and calmly before speaking or acting. You avoid risks whenever possible and like to plan ahead to ensure you can remain in control at all times. |

| 26–35 | You are somewhat less impulsive than most people and would sooner take time to reflect on the likely outcome of your actions before saying or doing anything risky. Most of the time you prefer to plan ahead to ensure you can remain in control at all times. |
| 36 – 55 | You can be impulsive on occasions, saying or doing things on the spur of the moment. Although you are not usually much of a risk taker there are times when you throw caution to the winds. |
| 56 – 80 | You are a highly impulsive individual who would sooner act on the spur of the moment than reflect on the possible consequences of your words or deeds. You enjoy taking risks and do not allow setbacks or mistakes to deter you. |

## Brain chemistry, risk and trust

The extent to which we are prepared to take a risk in our dealings with another person – for example by lending them money – depends, of course, on how much we trust them. When the level of trust is high enough even a dyed-in-the-wool Minimiser is likely to take a significant risk.

A number of chemicals in the brain are important for different aspects of social engagement and risk taking. Some of the most interesting studies involving risk taking and trust have involved oxytocin, a neuropeptide normally associated with childbirth and breastfeeding. 'When we induce the brain to release oxytocin in men and women they become more moral, they become more trustworthy, more generous, more compassionate, more empathic,' says Paul Zac of Claremont Graduate University.[13] He has found that when people interact with one another in a natural way around

65 per cent of them release oxytocin in the brain. Those whose brains have released oxytocin feel closer to the community and become more connected. Which raises the question that since oxytocin levels rise when we trust another person, would artificially raising its level promote greater risk taking by making people more trusting?

Research by Michael Kosfeld and his colleagues at the University of Zurich's Institute for Empirical Research in Economics suggests that this is exactly what happens.[14] In this study, almost half (45 per cent) of volunteers who had oxytocin administered by means of a nasal spray showed a maximum level of trust compared with only one in five (21 per cent) in a group given a placebo. 'Because we live in a sea of strangers the oxytocin system is a very fast on-off switch, so once it's turned on we engage,' explains Zac. 'But it turns off right away because the next person with whom we interact may not be safe or trustworthy.'[15] He cautions that attempts to increase trust by giving people a pill, or boosting their oxytocin levels in some other way, are doomed to failure. 'We still have control over the decisions we're making. Oxytocin is a very subtle signal. It motivates us to approach other people but that's about where it ends. It also depends on our genes, developmental history and even our current physiologic state.'

In the remaining chapters we will be exploring the ways in which risky decisions impact for good or ill in our everyday lives. Whether falling in love, going shopping, putting on weight, embarking on a career of crime, taking part in a riot or impulsively deciding to end it all.

# CHAPTER 8

# The Love Impulse –
# 'It Only Takes a Moment'

'There is a Lady sweet and kind,
Was never a face so pleased my mind;
I did but see her passing by,
And yet I love her till I die.'
Anon. 17th century

Humans are the sexiest animals on earth. The only species which, free from the constraints of reproductive necessity, engages in sexual activity whenever and wherever the opportunity arises. We regularly decide, usually on an impulse, whether or not we find another person sexually desirable. This typically split-second judgement – it can be reached in less than a hundredth of a second – proves remarkably resistant to change and likely to influence all our future interactions with that person. Yet while we may find it easy enough to say whether or not we are attracted to someone, it is usually far harder to explain exactly why they seem so sexually desirable. If asked what we see in them we usually fall back on a few stereotypes, such as our liking for blondes or preference for brunettes, our attraction to masculine six-pack men or penchant for the lean and brooding male. Saying we like, admire or desire certain people because these are the sort of people we like, admire or desire is a circular argument that does nothing to advance our understanding of this vitally important aspect of human behaviour.

## Instant desire and slow-burn romances

Research, as well as everyday experience, has identified two distinct types of sexual attraction. For some the attraction is instant. Eyes meet across a crowded room and people fall in love at first sight. In his autobiography, Bertrand Russell describes how, at the age of 17, he met and 'fell in love . . . at first sight' with 23-year-old Alys Pearsall Smith. Russell immediately decided he would, if possible, make her his wife. Noting in his diary that while he considered it unlikely she would remain unmarried 'until I grew up . . . I became increasingly determined that, if she did, I would ask her to marry me.'[1] When Russell came of age three years later, in May 1893, he began to pursue Alys with 'more than distant admiration'. But although his impulse to marry her had arisen at their very first meeting, it was not, at least according to his memoirs, one motivated by sexual desire. 'Although I was deeply in love, I felt no conscious desire for any physical relations. Indeed, I felt that my love had been desecrated when one night I had a sexual dream, in which it took a less ethereal form.'[2] What attracted him to Alys was not just her great physical beauty – she was later described as 'one of the most beautiful women it is possible to imagine' – but her personality, attitudes, beliefs and opinions: she was, Russell decided, a truly emancipated woman.

## Darwin's reasons for romance

On 2 October 1836, Charles Darwin stepped ashore from the *Beagle* on which he had served as the resident geologist during an expedition that had lasted almost five years. He was 27 years old and his head was full of radical ideas about evolution that he would, 23 years later, publish as *On the Origin of Species*. On his return to England, however, Darwin's mind was focused on a far more urgent personal matter:

whether or not to marry. Seated at his writing desk he divided a sheet of paper into two columns and wrote at the top: 'This is the question.' Then, under the heading 'Marry', he carefully set down what he regarded as the benefits of such a union: 'Children – (if it please God) – Constant companion, (& friend in old age) – object to be beloved and played with – better than a dog anyhow – Home, and someone to take care of house – Charms of music and female chit-chat. These things are good for one's health.' Under the heading 'Not Marry' he noted: 'Freedom to go where one liked – choice of Society and little of it. Conversation of clever men at clubs – Not forced to visit relatives, and to bend in every trifle – to have the expense and anxiety of children – perhaps quarrelling. Loss of time – cannot read in the evenings – flatness and idleness – Anxiety and responsibility – less money for books etc – if many children forced to gain one's bread – (But then it is very bad for one's health to work too much). Perhaps my wife won't like London; then the sentence is banishment and degradation into indolent idle fool.'[3] Having completed this very methodical, System R, 'cost-benefit' analysis he was left in no doubt as to his next move. Beneath the first column he wrote in a firm hand: 'Marry – Marry – Marry Q.E.D.' The following year he proposed to his cousin Emma Wedgwood. This he did diffidently, fearing himself so 'repellently plain' that she would turn down his proposal. Despite these fears he noted: 'Determined . . . to try my chance.'[4] The couple enjoyed a long and happy life together and had ten children, two of whom perished as infants. Their marriage was ended only by Charles's death in 1882, with Emma following 14 years later in 1896.

## Love and passion

Most of us would favour Russell's spontaneous System I impulse to Darwin's matter-of-fact and businesslike System R approach to marriage, despite its successful outcome. It should be noted, however, that while Darwin remained happily married, Russell's own private life was far less settled. He married and divorced four times, his last wife Edith being an educated and delightful woman whom I am privileged to have known.

Despite this, most of us believe the ideal is to fall passionately in love on an impulse. In a moment when, as James Joyce wrote in *Ulysses*: 'I asked him with my eyes to ask again yes, and then he asked me would I yes . . . and first I put my arms around him yes and drew him down to me so he could feel my breasts all perfume yes and his heart was going like mad and yes I said yes I will yes.'[5] The 'giddy high' this produces has been likened to a drug by New York psychiatrist Dr Michael Liebowitz. He argues that some people become so hooked they are transformed into 'love junkies': addicts who race from one relationship to the next in a desperate attempt to give themselves another romantic fix. Caring little about the object of their passions, such people are usually doomed to a series of disastrous relationships. Liebowitz believed that such frenetic romancing might be due to a lack of dopamine in the brain. Dopamine belongs to a class of substances known as catecholamines, which also include noradrenaline and adrenalin. These amines, together with another amine, serotonin (5-hydroxytryptamine), exert a huge influence on how we feel. Noradrenaline and dopamine influence motivation whilst serotonin influences mood. He prescribed a male 'love junkie' MAO (monoamine oxidase) inhibitor, an antidepressant which increases levels of dopamine in the brain. Within a few weeks his love-addicted patient had settled down to a more normal pattern of attraction. 'He no longer got so carried away by romance,' the doctor remarked. 'The frantic need to have somebody all the time vanished.'[6] But what is it about someone that triggers this surge of dopamine underlying the sexual impulse?

## Two theories of mutual attraction

Folk psychology offers two seemingly diametrically opposite theories. According to some, 'like attracts like', while others claim that 'opposites attract'. Surprisingly both turn out, at least to some extent, to be correct. In the early 1960s, Dr Carol Izard of Vanderbilt University administered a personality assessment to a group of newly arrived students. Six months later she measured their attraction for one another and found that those whose personalities were most similar had developed the closest friendships. Of particular importance were characteristics of a need for achievement, a desire to be self-sufficient and the extent to which an individual sought the company of others.[7] The message from this early research seemed clear. If two people share many attitudes and aspects of personality the chances are that they will be attracted to each other. The greater the overlap the more attractive they will find one another. Unfortunately for the advocates of what has been called the 'Narcissus hypothesis', after the handsome Greek youth who fell in love with his own reflection and drowned trying to reach it in a pool, subsequent research findings were not as straightforward as studies like these might suggest.

In his 1971 book *The Attraction Paradigm*, social psychologist Donn Byrne proposed that while shared attitudes appear to form the basis of attraction, the number of similar views matter less than the proportion of *similar* to *dissimilar attitude*. People appear willing to allow someone they like some disagreement with their own views without that individual becoming any less attractive. Once a critical number of differences have been reached, however, liking rapidly declines.[8] So consistent is this finding that one can express the probability of liking or disliking occurring in terms of the following mathematical formula:

$$Y = 5.44X + 6.62$$

This states that the best way of predicting the extent of liking (Y) between two people is to multiply the proportion of similar attitudes (X) by 5.44 and then add 6.62. Two people who share 50 per cent of the same views will score 9.34 (i.e. 0.5 × 5.44 + 6.62) on a 12-point scale. Should they hold 80 per cent of their views in common, then liking will rise to 10.97. The discovery that it is the *proportion* of shared attitude which is important might be accommodated within the 'Narcissus hypothesis', by replacing notions of the other person being a perfect carbon copy of oneself with a more fuzzy – though still recognisable – replica of one's own attitudes and opinions. For many social psychologists, however, this compromise seemed as unsatisfactory as the original theory. They argued that we are attracted to people who seem very much like ourselves because we assume that a shared outlook and attitude makes it more likely that the other person will like us in return, thereby reducing the risk that our advances will be rejected. In other words, we try to stack the cards in our favour as far as we can, by investing most of our time and effort on those who seem as much like us as possible. Support for this view came during the 1960's from the work of George Levinger of the University of Massachusetts and James Breedlove who found that when people are attracted to one another they tend to believe they have more in common than is the case. In one study, husbands and wives were asked to report their own attitudes as well as those of their spouse on a variety of topics. The researchers discovered a substantial discrepancy between the opinions which each held and those which their partner believed them to hold. They suggested that, in a close relationship, individuals tend to stress similarities of viewpoint in order to conceal or avoid sources of conflict.[9]

Another approach to understanding mutual attraction is based on the idea of each of us possessing two self-images. There is our Real Self, the person we see ourselves as being, and our Ideal Self, the sort of man or woman we would most like to be.

Attributes listed under the first heading might include the following: friendly; inclined to be bossy; generous; witty; sarcastic; outgoing; anxious; talkative and so on. The desired attributes of our Ideal Self, which we may feel we lack but would like to possess, might include generosity, assertiveness, sympathy and the ability to love. Our Real Self and our Ideal Self may be very similar to, or vastly different from, one another. It is not unusual to come across people whose Ideal Self has little in common with the reality of their lives. This can lead to great unhappiness and frustration, negative emotions that can only be resolved in one of three ways. Firstly by changing one's behaviour so that it comes closer to the ideal. Secondly by changing one's Ideal Self to make it more accurately resemble one's real aspirations and achievements. Thirdly by striking up a relationship with a person who seems to embody most if not all of those qualities and successes that go into one's Ideal Self. We will find such a person psychologically, although not necessarily physically, attractive. The extent of this attraction will vary according to the extent the other person 'acts out' or embodies those aspects of personality we would like to possess but believe, rightly or wrongly, are beyond our ability to exhibit. In this way someone who lacks confidence and is indecisive may be strongly attracted to one who is overflowing with self-confidence and makes swift, bold, decisions. By impulsively forming such an attachment, which may be close and personal or emotionally unrequited, we are able to satisfy, if only precariously, the goals of our Ideal Self. The people who fulfil this role in our life may be sexual partners or merely friends, colleagues, employers, teachers or therapists we know personally. Equally they may be total strangers known exclusively through the media. Such role models include royalty or political leaders; industrialists or union leaders; pop stars, sports stars, film, stage or TV stars. They may be religious teachers or mystics; philosophers or painters; soldiers or scientists. They may even be mass murderers, terrorists or torturers, for no matter how ugly their deeds, someone, somewhere is almost certain to see them as essential components of their ideal self. There are, after all, still shrines to Stalin, Mao, Hitler and Pol Pot at which the faithful continue to revere the memory of these genocidal despots.

## An eye for beauty

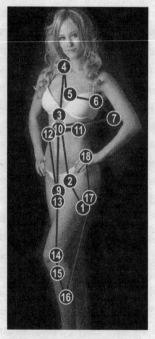

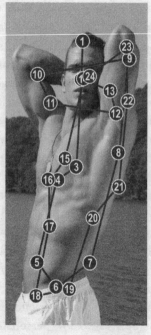

Figure 1:
Man viewing woman

Figure 2:
Woman viewing man

In my laboratory we use eye-tracking technology to examine how men and women look at one another's bodies. In the figures above, the numbers in the circles indicate the direction of gaze while the size of the circles shows how long – in milliseconds – the eyes fixated on that spot.

The illustrations above show that, at least under laboratory conditions, men and women look at the semi-naked bodies of members of the opposite sex in very different ways. Figure 1 shows the direction of gaze of a typical, single heterosexual man viewing a bikini-clad woman. He started at her thighs and buttocks (1), and then moved up to her groin (2), breasts (3) and neck (4). His eyes travelled back to her breasts (5 and 6), and up and down her abdomen (8–12) and legs (13–17). He concluded at the point where her left hand points downward to her groin (18). The woman, by contrast, began her inspection at the man's face (1), returning there on three subsequent occasions. Next, her eyes travelled up and down his chest (3 and 4) and abdomen (5), halting just above his groin (6). She then examined his chest muscles and left arm (7–9), his right arm (10 and 11), his left arm-pit (12), and then his face again (13). She repeated her examination of his torso (15–17), stopping, as before, just above the pubic region (18). Finally, her eyes moved back up his left arm (20–3) before ending where she began, at his face (24). When the individuals being viewed are fully clothed a different pattern of gaze occurs with more attention being directed towards the face.

## Beautiful faces are symmetrical faces

The crucial importance of faces in triggering the (not always compatible) impulses of love and lust was demonstrated in an intriguing series of studies by Adrian Furnham and his colleagues at the University of London. They asked young males to judge photographs of women in terms of their attractiveness, sexiness, fertility, healthiness and the probability of their being pregnant. The researchers found that an attractive face led to the women being judged healthier, sexier, more fertile and more attractive. 'As facial features can also give cues to age and, as a result, reproductive capability,' the authors comment, 'choosing a facially attractive female would result in choosing a female who was fertile, healthy and whose genes were of good quality.'10 Few of us have faces that are classically symmetrical. More often than not, the features are misaligned with slightly different characteristics making one side of our face unlike the other. If you study your own features in a mirror you will see that, other differences aside, your eyes slant slightly downward from either left to right or the other way around while your mouth does the same. Your 'good side', from a photographic viewpoint, is the side where the distance between the corner of the mouth and the edge of the eye is at its widest since this will confer the right perspective to the image.11

The final clue we pick up from the face, and one capable of significantly, if subconsciously, influencing our judgements when seeking a life partner, is how healthy they are. Ethnographic surveys going back more than 50 years[12] have demonstrated the importance, in every culture studied, of clear eyes, full lips and a fresh complexion. Nowhere in the world were bloodshot or yellow eyes, chapped or scarred lips, or heavily blemished skin seen as desirable. This finding applies as equally to men as to women.[13]The one thing that beautiful faces have in common is their symmetry. In classical Greece this was considered the key to true beauty and in the 5th century BCE the Greek artist Polyclitus wrote a famous essay extolling its virtue. What was true then remains true to this day. The more symmetrical their face the more likely a person is to be regarded as attractive, desirable, healthy and fertile. The less symmetrical, the more likely they are to be seen as having poor health and lower fertility.[14] What applies to the face also applies, in equal measure, to the body.

## The body beautiful

In ancient Greece the ideal male body combined muscular arms, strong shoulders, powerful abdominals and a tapered 'V' shape that identified him as able to both protect and provide.[15] The type of physique which, thousands of years later, most women find particularly attractive.[16] For the Greeks, and subsequent generations, the curvaceous figure of the goddess Aphrodite represents the classical idea of the perfect female form. 'Greek art reflects the physical features of a man and a woman deemed most likely to be ideal breeders and so, in Darwinian terms, to increase their own, and mankind's, chances of survival,' comments neuroscientist Charlie Rose. 'Females were attracted to males whose physique indicated muscular strength and endurance, cardiorespiratory power and enhanced motor skills essential for hunting and fighting – the two skills on which survival depended during most of man's history. Not only would "fit" males increase the chance of survival of their mates and mates' children, it was also likely that their adaptive traits would be inherited by their offspring. This made "fit" males more competitive against weaker males in mating success.'[17]

As in classical Greece, features of the female form associated with a high level of reproductive success[18], such as a curvaceous figure created by a low waist-to-hip ratio, are those most likely to trigger the impulse of sexual desire. 'It is the fundamental assumption of all evolution-based theories of human mate selection that physical attractiveness is largely a reflection of reliable cues to a woman's reproductive success,' says Dr Devendra Singh of the University of Texas. 'A man can increase his reproductive success by choosing a woman who is receptive, highly fecund, and has characteristics suggestive of being a successful mother. A woman can increase her reproductive success by choosing a high-status man who controls resources and, hence, can provide material security to successfully raise her offspring.'[19] The reproductive value of a man, as a rule, can be easily assessed because high status is usually achieved through competition with other members of the social and economic hierarchy.

## 0.7 – The Fertility Factor?

In the absence of any direct signals of ovulation or fertility, the man is forced to use indirect cues such as physical attractiveness to assess the reproductive value of the woman. The most reliable of those fertility cues is the waist-to-hip ratio (WHR) which, Dr Devendra Singh contends, is the key to being regarded by males as fertile and fanciable. The waist-to-hip ratio[20] is calculated by dividing the circumference at the waist by the circumference at the hip. A 60cm waist and 83cm hip circumference, for example, produces a WHR of 0.7. This ratio, which is highly correlated with attractiveness, not only predicts a woman's vulnerability to disease but also her fertility. Researchers have found, for example, that despite changes in height and weight, the 'hourglass' body shape of Miss America contestants remained the same down the decades.[21] Between 1920 and the 1980s, every winner had a WHR between 0.72 and 0.69 while that of *Playboy* models was between 0.71 and 0.68. During the 20th century such beauty icons as Marilyn Monroe, Sophia Loren, Twiggy and Kate Moss all shared a WHR of about 0.7.[22]

There are, however, significant cultural differences. In 2001, anthropologists Frank Marlowe and Adam Wetsman showed drawings of women, which varied only in their WHR, to groups of Hadza, Tanzanian hunter-gatherers. When comparing their preferences with those of Americans, they found that although the 0.7 WHR held true for US males the Hadza man favoured much higher ratios, showing their ideal to be far plumper females. Marlow and Wetsman speculated that the more subsistence-orientated a society and the more demanding work done by women, the greater the preferences for a fuller figure. 'Among foragers,' they noted, 'thinness probably indicates poorer health.'[23]

A similar argument has been advanced for the Rubenesque plumpness favoured by wealthy males during the 16th and 17th centuries. During that period having a generously proportioned, fair-skinned wife was a status symbol since it meant she was a lady of leisure rather than a worker in the fields, who would possess a leaner and more sun-tanned appearance. When hosting afternoon tea parties became fashionable among the growing middle classes of the 18th century, a lady would expose her forearm to the others as she raised a bowl[24] containing the – at that time extremely costly – beverage to her lips. Like the expensive tea and the valuable beeswax candles that illuminated the elegantly furnished living room, without producing the unpleasant smoke created by the cheaper tallow candles used by the poor, the pale skin of a feminine forearm was a symbol of wealth, leisure and luxury.

## Why arousal leads to attraction

White-knuckle rides may not be everyone's idea of a romantic tryst, but perhaps they should be. I first studied the potent relationship between roller-coasters and sexual arousal some 30 years ago. In my research I equipped couples with heart-rate monitors as they were about to board a ride and divided them into two groups.

Couples in the first group sat side by side while those in the second were separated and placed next to strangers. The results were surprising. Those seated beside their partner not only experienced higher heart rates throughout the ride, but also reported heightened physical attraction to one another. They were far more physically demonstrative on leaving the ride, holding hands, cuddling and embracing. The couples who had been separated experienced slightly lower heart rates and were less physically attracted and close after the ride. The greater the thrill experienced in the presence of one's partner the greater the impact on emotional and sexual excitement. The findings of this early study, which involved a total of 40 couples, were subsequently confirmed by a great many other researchers.[25] In a study published in the *Archives of Sexual Behavior*, for example, Professor Cindy Meston of the University of Texas Sexual Psychophysiology Laboratory and Dr Penny Frohlich asked individuals who were either waiting to start or had just completed a roller-coaster ride to rate a photograph of an averagely good-looking member of the opposite sex on his or her attractiveness and dating desirability. As they predicted, those who had just ridden the roller-coaster rated the person in the picture higher on both counts than did those waiting to take the ride.[26] So why does it happen? What links the thrill of a high-speed modern roller-coaster ride with sexual attraction and arousal?

## Adrenalin buzz and the 'misattribution effect'

The key lies in the surge of the powerful 'fight-or-flight' hormone adrenalin which is triggered by the speed, the turns and above all the sudden increase in G-force to which people are subjected on a modern roller-coaster. The more thrilling and nerve-tingling the ride, the greater the adrenalin release and the bigger the sexual buzz.

One of the earliest studies of the relationship between increased levels of adrenalin and sexual arousal was conducted more than 40 years ago. In 1974 two researchers[27] positioned an attractive female interviewer in the centre of either a sturdy bridge or a far flimsier and less stable walkway. In each case she stopped men crossing the bridge and asked them to complete a survey. After they had done so, she supplied her name and phone number so they could contact her if they had any questions regarding the survey questions. It was found that the men interviewed on the unsteady bridge were significantly more likely to call her than those interviewed on the sturdy bridge. The researchers suggested that their heightened sense of anxiety and increased adrenalin production produced greater levels of sexual excitement and interest in the attractive female.

In 1989 scientists[28] conducted a somewhat similar piece of research, this time in a cinema. They studied sexual attraction and arousal among couples who had just seen one of two types of film. One was a nail-biting, highly arousing suspense movie, the other a far less arousing film. The researchers found that the couples who had viewed the suspense film were far more likely to be touching, holding, embracing and kissing one another than those leaving the less exciting movie.

All these results demonstrate what psychologists call a 'misattribution effect'. This occurs when emotional, physical and sexual arousal generated by one situation – the roller-coaster ride, the flimsy bridge or the scary film – is mistaken for sexual excitement provoked by another person.

## Personality and the sexual impulse

In his 1772 *Essays on Physiognomy*, the Swiss poet Johann Caspar Lavater gives a detailed description of how to relate facial features to personality traits. His book includes such assertions as: 'The nearer the eyebrows are to the eyes, the more earnest, deep, and firm the character.'[29] Although his ideas seem ludicrous today, with their echoes of phrenology, there is good evidence that attractive people enjoy outcomes superior to those of unattractive people.[30]

In an early study, Drs Jerry Wiggins, Nancy Wiggins and Judith Conger from the University of Illinois decided to test male 'locker-room' folklore that there are 'leg men', 'breast men' and 'bottom men'.[31] They administered a personality test to a group of males and then asked them to rate different body parts on the silhouettes of naked females in terms of attraction. An example of the illustrations from their original paper is shown in figure 3.

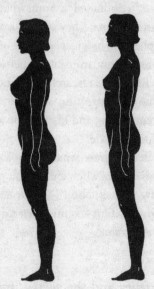

*Figure 3: Silhouettes of two naked females*

Here, in summary, is what they found.

**Large breasts**: Men with this preference dated frequently, enjoyed typical 'masculine' interests, followed sport and had an above-average sex drive. They enjoyed showing off and being the centre of attention. They preferred relationships that allowed them a great deal of freedom and objected to having their independence challenged. They were most likely to read *Playboy* magazine!

**Small breasts:** This was the choice of men who drank little alcohol, experienced bouts of mild depression and sought warm, empathic relationships. Somewhat conformist in their social outlook, they had a strong need for achievement and worked hard to make a success of their careers, persevering despite setbacks and difficulties.

**Large buttocks:** This preference was related to a desire for order and organisation in life. Males who favoured this part of the anatomy sought the views of others before taking decisions and liked to have support for their actions. They were dependent in their relationships and tended to blame themselves when things went wrong.

**Small buttocks:** This preference was associated with the ability to persevere in complex or demanding tasks and a reluctance to accept the blame when things went wrong. These men had little desire to be the centre of attention, and were unlikely to take much interest in sports.

**Short legs:** This preference was linked to a liking for the company of others and to social situations in which they were the focus of attention. They had a strong desire to be helpful and the need to feel wanted, trusted and liked by others.

**Long legs:** Men who favoured long legs were non-aggressive and blamed themselves when things went wrong, even when the mistake was not really their fault. They were deep thinkers and more introspective than most, preferring to live life at a relatively slow pace and showing little interest in business matters.

Although I find some of these results hard to swallow, there is no doubt that preferences for different body types is in some way linked to aspects of our personality.

A similar study by Dr Sally Beck of Butler University[32], this time using silhouettes of males, showed that female preferences also provide clues about the personality of those who make the choices. The women who took part in her study overwhelmingly favoured men with small buttocks. Although moderately large chests were favoured, there was less preference for very large ones. The length of a man's legs made no difference to his attractiveness. Some of her other findings were:

**Large thickset man:** Women who prefer this body shape in the main usually enjoy sport and physical activity and regard themselves as less traditionally feminine.

**Large chest, small buttocks:** This choice is most often made by women with a professional or academic background.

**Small chest, moderate-sized body:** Women with a more traditional view of the female role are most attracted by this outline. It is most likely to be chosen by women from higher social and economic backgrounds.

As can be seen, the research suggests a far less specific link between particular parts of the body and traits of personality. In general, the physical characteristics women desire in men differ greatly from the masculine stereotype. Especially appealing, according to writer Rosalind Coward, is 'sexual surprise'[33]: that is, the discovery of female characteristics within an otherwise 'masculine' appearance. These features include long eyelashes, pretty faces and soft hair on muscled arms. Trim, rounded but firm male bottoms are also liked – the kind of bottom that could just as easily be seen on the woman. The most disliked features included a drooping moustache, wiry hair protruding from a tightly buttoned denim shirt and bulging muscles.

## Love, lust and the evolutionary imperative

The psychology of love and lust are both extremely complex and fundamentally simple. The complexity arises from the layers of social and cultural traditions that have grown up around sexuality and mate selection. Traditions strongly shaped and profoundly influenced by metaphor, poetry, music, song and, in recent years, marketing. Today, love and marriage are big business, as, too, is lust, whose demands are catered for by services ranging from sex shops to the world's billion-dollar porn industry.[34]

At the same time the impulse to fall in love, or at least to make love, with someone can be regarded as fulfilling the demands not of culture but biology. Because the evolutionary histories of men and women are so different, each has evolved psychological mechanisms designed to solve problems specific to their sex. The evolutionary challenge facing women was to find a mate physically capable of protecting herself and their children while providing the resources essential for survival.[35] Although these days such strength is more likely to be derived from financial clout and social status than physical prowess, old evolutionary habits die hard! In addition, since human babies – unlike those of every other species – remain dependent on their parents for a lengthy period, she will also ask herself: Is he the kind of man to stay around and help me raise our young? Heterosexual men will want to know whether their chosen partner is healthy and fertile in order to ensure their genes can live on through the children they produce.

In summary, although concepts of beauty and attractiveness are strongly influenced by the media, pop music, the internet, celebrities, fashion writers, parents and friends, to name but a few, research suggests that these preferences are not arbitrary. Rather, they reflect evolutionary cues to fertility, health and, more generally, that person's perceived value as a breeding partner.

## Scarcity enhances attraction

The availability of potential partners is yet another factor that can influence sexual attraction and help trigger the love impulse. The 'scarcity' effect on physical desire was first investigated by Dr James Pennebaker of the University of Virginia psychology department who was intrigued by a line in songwriter Baker Knight 1970s hit: 'Don't All the Girls Get Prettier at Closing Time'. This read: 'Ain't it funny, ain't it strange, the way a man's opinions change, when he starts to face that lonely night.' Deciding that this was a testable theory, Pennebaker[36] sent research assistants into singles bars at hourly intervals between 9.30 p.m., and closing time. They focused on men and women who were alone and appeared 'fairly sober'. Adopting an approach suggested by the lyric 'If I could await them on a scale of one to ten, looking for a nine but eight could fit right in', the researchers asked people to use this rating on members of the opposite sex. The results provided strong support for Baker Knight's lyrics. As a closing time crept closer and closer, people without partners started to see remaining members of the opposite sex as increasingly attractive. A man or woman rated four when the bar was full might be rated six as numbers dwindled and those who remained looked like facing a 'lonely night'.

## Myths of partner choice

Before leaving the love impulse, I need to correct some of the myths about differences between men and women when it comes to sexual behaviour. There is a widely held view, for example, that men enjoy sex more than women; select their partners solely on sexual appeal; seek out more sexual partners; experience more orgasms; engage in casual sex more frequently; are less choosy about romantic partners and think about sex every seven seconds. This view of the love impulse is summed up by the cynical aphorism: 'Marriage is the price men pay for sex. Sex is the price women pay for marriage.'

'Because this framework is popular among both psychologists and the general public, it contributes to face-value acceptance of sex related gender differences,' say Terri Conley and her colleagues at the University of Michigan. 'Gender differences are often viewed as supporting biological, genetic, or evolutionary accounts of mating; however, upon further empirical scrutiny, these gender differences are either not what they seem, narrow considerably, or in some cases, are completely eliminated.'[37]

In their research they investigated the six most widely accepted differences between men and women when it comes to sex.

### 1. Men seek sexy women and women high-status males.
**False**: While this notion is often supported by anecdotal examples of high-status older males with young and beautiful trophy wives, research shows conventional wisdom to be wrong. In both initial speed-dating encounters and on follow-up a month later, attractiveness and status were found to be of equal importance to both men and women.

### 2. Men want casual sex more than women.
**False:** A greater willingness to engage in casual sex is one of the most widely documented differences between men and women.[38] This was apparently confirmed by a major 1989 study[39] in which male and female research-assistant confederates invited participants to have casual sex. While 70 per cent of the men approached by women agreed, none of the women approached by male researchers did so. However, Conley[40] found that these seemingly highly significant differences disappeared when participants

considered sexual offers from very attractive or very unattractive but famous individuals. 'Likewise, women and men were equally likely to accept offers of casual sex from close friends whom they perceived to have high sexual capabilities (i.e., whom they thought would "be a great lover" and would provide them with "a positive sexual experience"),' says Conley. Her conclusion? Men, as a whole, are no more motivated to engage in casual sex than are women.

### 3. Women have fewer orgasms than men.
**Partly true:** The researchers found this was true to only a limited extent. The difference diminished substantially when the woman was in a long-term relationship and could disappear completely with some of the varieties of sexual activities within such relationships. 'Biological differences appear to have little to do with women's potential for orgasm,' says Conley. 'Instead, the sexual practices performed play a significant role in narrowing the orgasm gap.'

### 4. Women are choosier about their sexual partners than men.
**True:** But mainly because men are more likely to chase women than the other way around. In a study where women approached men, they started to behave more like men, by becoming less choosy. The men being approached, on the other hand, became *more* choosy.[41]

### 5. Women desire and actually have fewer sexual partners than men.
**False:** In a fascinating study[42], researchers divided participants into two groups and attached one to what they claimed was a powerful lie detector, a machine that could distinguish truth from falsehood with 100 per cent accuracy. As anticipated, men in the group which had *not* been wired up to the infallible lie detector claimed to have

had more sexual partners than did the women. However, those males attached to the bogus polygraph made no such claims and the supposed difference between the sexes vanished.

## 6. Men think about sex more than women.
**True:** But only because they think more about all their physical needs, including food and sleep, more often than women. In summary, researchers found that perceived differences between the sexes arise from such causes as stigma against women for expressing sexual desires; women's socialisation to attend to other's needs rather than their own and 'a double standard that dictates (different sets of) appropriate sexual behaviours for men and women.'[43]

In his monumental *À la recherche du temps perdu* (*In Search of Lost Time*) French author Marcel Proust clearly describes the emotional conflicts so often created by the love impulse, whether sparked by heterosexual or homosexual desire.

'What we suppose to be our love or our jealousy is never a single, continuous or indivisible passion,' he wrote. 'It is composed of an infinity of successive loves, of different jealousies, each of which is ephemeral although by their uninterrupted multiplicity they give us the impression of continuity, the illusion of unity.'

In this chapter I have examined some of the frequently over-looked aspects of love, sex and desire. In the next we will explore another, equally addictive type of impulse, the one which results in the seventh of the seven deadly sins: gluttony.

# CHAPTER 9

# The Overeating Impulse – Digging Our Graves with Our Teeth

'I can resist anything but temptation.' Oscar Wilde,
*Lady Windermere's Fan*

We make some 200 decisions about what, where, when and how much to eat each day[1], the majority on an impulse. These are mindless food choices, which, more than anything else, helps explain why so many people are risking their health and shortening their lives by piling on the pounds.

Obesity is nothing new. The Hohle Fels Venus, carved from mammoth ivory and dug up in Germany in 2008, depicts a short, squat woman with pendulous breasts, broad shoulders, wide hips and a bulging belly. Suitable models, both male and female, can be seen in the towns and cities of most developed nations today. In fact this overweight lady is more than 35,000 years old. During the 18th century visitors flocked to gawp at Daniel Lambert, who was 5 foot 11 inches tall and weighed 52 stone. He earned his living by exhibiting his monstrous bulk around Britain. Yet, even 60 years ago, Billy Bunters and Fatty Arbuckles were thin on the ground, especially among the working classes. For the vast majority of blue-collar workers in England, Europe and America, the problem was an absence rather than an excess of wholesome, nourishing food. In the industrialised world most of the population had, like Cassius in Shakespeare's *Julius Caesar*, a 'lean and hungry look'. Today in the developed world being too thin is as

unusual as being too fat was in our grandparents' day. The impulse
to guzzle chocolate, order an extra side dish of French fries or tell
the fast-food assistant to 'supersize me' has resulted in the one
billion overweight adults on the planet, 300 million of whom are
obese.[2]

Obesity is defined as a Body Mass Index (BMI)[3] of 30 or higher,
while a BMI of between 25 and 30 marks a person out as being
overweight but not obese. I am using the BMI here because it is
so widely employed and recognised. When applied to individuals
rather than groups it is, however, the statistical equivalent of snake
oil, for reasons I give in the chapter notes at the end of the book.[4]
Obesity is the developed world's most urgent public health
problem. An epidemic which will, doctors predict, produce an
additional 6 to 9 million cases of diabetes, 6 to 7 million cases of
heart disease and stroke, between 492,000 and 669,000 additional
cases of cancer, and the loss of up to 55 million life years in the
USA and UK combined. Within 20 years, they say: 'The combined
medical costs associated with treatment of these preventable
diseases are estimated to increase by $48–$66 billion a year in the
USA and by £1.9–£2 billion a year in the UK.'[5]

Studies have shown that 20 per cent of people born between
1966 and 1985 became obese by their early twenties, a milestone
not reached by their parents until their thirties or by their grandparents until their forties or fifties. Overweight children run a far
higher risk of suffering from serious medical conditions, including
digestive disorders, heart and circulatory illnesses, respiratory
problems, depression and other forms of mental illness. In the
USA some 30 per cent of boys and 40 per cent of girls are at risk
of being diagnosed with Type 2 diabetes at some point in their
lives. For the first time in history obese children may not outlive
their parents.[6] But why is the impulse to eat chocolates so much
stronger and more rewarding than to consume carrots? Why do

artery-clogging cream buns taste so much more delicious than health-giving broccoli?

## Junk food and jungle survival

While the food industry must take much of the blame for a near epidemic of bulging waists and shortened life spans, it is also important to recognise the psychological reasons why we relish the foods that pile on the pounds. What is going on inside our head when we decide whether or not to indulge in a second helping of French fries or another milk sundae? Impulse eating is triggered by what psychologists call 'palatable food'. To understand more clearly what this term means we must journey far back in history to a time when our earliest ancestors roamed the earth. Just as sexual desire and mate selection, which we looked at in the previous chapter, are strongly influenced by primitive survival needs, so too are present-day eating habits still largely governed by our evolutionary past. Imagine yourself as a hunter-gatherer in Africa's Rift Valley some 200,000 years ago. You wear little, perhaps a scanty loincloth, carry a primitive club and will not evolve into the direct ancestor of modern man, *Homo sapiens* or 'wise man', for another 150,000 years. Each day is a struggle for survival. Your life, like that described by Thomas Hobbes in *Leviathan*, is 'poor, nasty, brutish, and short'. Only a quarter of your companions will see 40. Every daylight hour is devoted to scavenging for food, hunting down your next meal or resting after a feast. Bursts of energy-sapping effort are punctuated by periods of satiated bliss. After you and your companions have caught, killed and consumed an animal, and gorged yourselves on freshly picked fruit or sweet, sun-ripened berries, you relax with a full belly under the hot African sun and energy levels are restored. Faced with this endlessly repeated cycle of energy-demanding hunting and energy-delivering feasting, the human brain evolves an acute sensitivity to the tastes and textures of those foods that provide the maximum amounts of energy. This means things that are high in sugars and fats. In time not only does our brain become skilled at separating these foodstuffs from those providing less of a quick energy boost but it also develops a taste for them.[7] As a result, the scarcity and unpredictable availability of food during mankind's early evolution made all foods high in sugar and fat highly prized and sought-after commodities.[8]

## Our need to feed

Two systems control our need to eat: the physiological and the psychological. The former produces what is called 'homeostatic hunger' and the latter 'hedonic hunger'.[9] Physiological hunger is guided by signals sent from the gut, blood and brain: critically, the hypothalamus, which constantly monitors levels of glucose, amino acids and fatty acids. People experience physical hunger when low blood glucose levels lead to a loss of energy and feelings of discomfort. By consuming food they restore their homeostatic balance. While homeostatic hunger is primarily controlled by the hypothalamus and its connections with other brain regions, hedonic (or psychological) hunger, and the pleasure that we experience when eating palatable food, arises from a chain of events that start in the middle regions of the brain and extend outward to the frontal cortical brain regions. Critically, these are the same brain regions responsible for impulses.[10] A desire to savour some delicious treat creates a perfect storm of reward-producing chemicals (including dopamine, serotonin, opioids and cannabinoids) in the brain that overwhelm our ability to resist. It is this increase in these pleasure-providing endorphins, and the interactions between them, that makes fat-mixed-with-sugar taste so delicious.[11][12] Palatable foods, such as crisps, chocolates, cakes, burgers, sausages and pork pies, activate this cascade of hormones and neurotransmitters in the brainstem, a rush of activity that can be triggered long before you take your first mouthful by the mere sight, or even the thought, of a tasty treat.[13] Their release also gives rise to 'comfort eating' as they dull us, or at least distract us, from painful experiences. Doctors, for example, will often place a little sugar solution into a newborn baby's mouth before taking a blood sample. The sweet taste causes a surge of endogenous opioids and dopamine which prevent the baby from being distressed by the experience. By stimulating the release of potent reward hormones and neurotransmitters the food evokes a strong compulsion to have more of the same.[14] By overriding any sensations of 'fullness' it makes it far more difficult to know when to stop. This is the main reason why so many find it so hard to overcome their craving for the high-fat and high-sugar foods most likely to result in weight gain. For the same reason we cannot just conceal our favourite treats and hope to forget them. We will always remember the chocolate cake at the back of the larder or the biscuits stored out of sight above the fridge and know how good they will taste.

## Impulsive overeating

A large and fast-growing volume of research demonstrates that the relationship between impulsivity and overeating is bi-directional.[15] Which is to say, people who are overweight are often impulsive[16] and impulsivity often leads to weight gain: this applies to children as well as to adults.[17] That impulsive people are significantly more sensitive to images of high-calorie foods than less impulsive individuals, for example, has been demonstrated using magnetic resonance imaging. In one study healthy individuals were shown a series of images of delicious versus boring 'neutral' foods while having their brain scanned. The more impulsive they were, the greater the activity seen in areas responsible for reward.[18] Unfortunately for our waistline, the kind of foods most likely to be eaten on impulse are not just delicious. They are also addictive and are leading to a rapid rise in people we might fairly describe as fast-food junkies. 'In 1970, Americans spent about $6 billion on fast food; in 2000, they spent more than $10 billion,' says author Eric Schlosser in *Fast Food Nation*. 'Americans now spend more on fast food than on higher education, personal computers, computer software or new cars. They spend more on fast food than on movies, books, magazines, newspapers, videos, and recorded music – combined.'[19]

The fact is, that once tasted, foods high in sugars and fats rapidly create a desire to repeat the enjoyable experience as swiftly as possible. A wealth of studies using functional magnetic resonance technology to record blood flow in the brain provide compelling evidence that junk food leads to a compulsion to overeat. The fact we use the word 'craving' to describe an intense desire for both some types of food and drugs of abuse is no mere accident of language. Research shows that the impulse to overeat is triggered by the same brain systems that lead some people to smoke dope or shoot up on crack cocaine. One of the most controversial findings to have emerged from recent research is that junk food, when consumed to excess, risks transforming people into junkies. Most of the research in this area has, for obvious ethical reasons, been undertaken on laboratory animals and one must always exercise caution when applying results from animal research to humans. In studies of weight-loss drugs, for example, rats lost up to 30 per

cent of their weight while humans taking an identical drug lost
less than 5 per cent. 'You can't mimic completely human behaviour,' says Dr Gene-Jack Wang, chair of the medical department
at the US Department of Energy's Brookhaven National Laboratory
in Upton, New York, 'but [animal studies] can give you a clue
about what can happen in humans.'[20]

One such piece of research, conducted by Drs Paul Kenny and
Paul Johnson of the Scripps Research Institute in Jupiter, Florida[21]
suggested that high-fat, high-calorie foods affect the brain in much
the same way as cocaine and heroin. In their study they divided
rats into three groups and fed one group only standard rat food
for the next 40 days. The second set of rats were allowed to gorge
on such tasty but fattening human foods as sausages, bacon and
cheesecake. They were, however, only fed for one hour each day.
The third group enjoyed an identical high-fat diet which was available to them around the clock. As expected, those rats who binged
on unlimited junk food rapidly became obese. More significantly,
their brains also changed. Using electrodes surgically implanted
in the animals' brains to monitor electrical activity, Kenny and
Johnson found rats in the third group slowly built up a tolerance
to the pleasure the food provided. Then, like drug addicts, they
had to consume ever greater amounts of food to enjoy the same
high. Not only were they eating compulsively but they continued
doing so even in the presence of pain. When an electric shock was
applied to their feet in the presence of the food, rats in the first
two groups simply stopped eating. The obese group, by contrast,
ignored the painful shocks and continued to gorge themselves.
'Their attention was solely focused on consuming food,' explains
Paul Kenny. At this point their compulsion to food closely resem-
bled drug addiction. 'People know intuitively that there's more to
[overeating] than just willpower,' he says. 'There's a system in the
brain that's been turned on or over-activated, and that's driving

[overeating] at some subconscious level.' The brain changes observed in these obese rats were very much like those found in drug addicts and were produced in the same way: through a significant decrease in the levels of dopamine, a neurotransmitter involved in the brain's reward.

Despite the fact that far fewer studies have been conducted using human subjects, many lines of evidence support the view that a common brain mechanism exists for an addiction to both hard drugs and junk food. Heroin, for example, stimulates the production of opioids while cocaine elicits dopamine. Activity in the orbitofrontal cortex, that portion of the brain above the eyes, is linked to cravings for both alcohol and cocaine. But there may well be another reason why some people, and especially some children, have such difficulty in knowing when to stop eating. And it's all down to the absence of a type of bacteria in the stomach which modern living and modern medical practice seems to be driving into extinction.

## The bacteria on which our life depends

Our life depends on the vast armies of bacteria – they outnumber human cells by ten to one – with whom, for most of the time, we enjoy a symbiotic relationship. Living on our skin, genital areas, mouth, stomach and intestines they perform a wide range of essential tasks. These include helping break down food into the nutrients we need to provide energy and repair tissues. One such bacterial species, *Helicobacter pylori*, lives in the stomach, where it helps regulate acid. During the 1980s two Australian physicians, Barry Marshall and Robin Warren, identified it as the causative agent of peptic ulcers. Although their theory was regarded as highly controversial at the time, treating peptic ulcers with antibiotics today has led to a 50 per cent reduction in ulcers induced by *H. pylori.*

But *Helicobacter pylori* also plays a crucial role in regulating our appetite. It has long been known that the stomach produces two hormones responsible for controlling hunger. One of them, ghrelin (from a word meaning 'to grow'), stimulates our desire to eat while the second, leptin (produced by fat tissue), tells us when to stop eating. 'When you wake up in the morning and you're hungry it's because your ghrelin levels are high,' says Martin Blaser, Professor of Internal Medicine and Microbiology at New York University. 'The hormone is telling you to eat. After you eat breakfast, ghrelin goes down.'[22] Researchers have discovered that *H. pylori* plays a vital role in reducing ghrelin levels after a meal. In its absence people continue to feel hungry and so carry on eating. Unfortunately *H. pylori*, which (in its present form) humans have been carrying around inside them for at least 60,000 years, is rapidly becoming an endangered species. Two or three generations ago some 80 per cent of Americans had these important bacteria in their stomach. Today fewer than 6 per cent of American children have been found to harbour them. One of the reasons for this decline, according to Martin Blaser, is the widespread use of antibiotics. 'We now are more than 60 years into the antibiotic era, and in developed countries, children regularly receive multiple courses of antibiotics for various ailments, especially otitis media [ear infection]. If each course of antibiotics eradicated H. pylori in 5–20 per cent of cases, the cumulative effect of childhood antibiotic regimens would remove a substantial proportion of colonisations.'[23] One of the key factors in childhood obesity may, then, turn out to be changes – partly due to antibiotic use – to their intestinal bacteria. They continue to eat not because they are greedy but because they still feel hungry. The impulse results from the decline of a vital bacterial helper in their stomachs rather than a lack of self-control or willpower.

## The diet impulse

Confronted by trousers and skirts too tight to button and worried about the rolls of socially unsightly fat around one's waist, the obvious answer would seem to be to change one's eating habits dramatically. To reduce the amount of food eaten, to cut out high-fat, high-carbohydrate or high-protein foods (depending on which of the latest dieting crazes most appeals) and then watch the pounds fall away. Unfortunately, however worthy the resolution, the outcome is likely to be disappointing. Most people, research suggests, cycle wildly between feast and famine. Between bingeing and starving, with weight coming off slowly and then piling back on rapidly. The amount of self-control involved in serious dieting is so biologically costly that it is extremely hard to sustain for any length of time. This is especially true in most large towns and cities where fast-food restaurants can be found in almost every major street, often duplicated many times over.[24] Further, yo-yo dieting may actually do more harm than good. Recent studies have raised the worrying suggestion that such feast-and-famine dieting could even lead to brain damage. As I explained, foods high in fat and sugar bring about changes in brain function, similar to those produced by drug abuse, by triggering an increase in both dopaminergic and opioidergic activity. The euphoria generated by this combination of 'reward hormones' can be compared with that produced by injecting or snorting one of the most potently addictive street drugs, speed balls. This cocktail of heroin and either morphine or cocaine can cause a strong physical dependence and withdrawal symptoms.[25] When people first flood their brain with dopamine and opioids by overeating junk food, then reduce them through dieting, only to trigger another surge as they slip into a pattern of yo-yo eating and dieting, they risk saturating and depleting their dopamine receptors. As a result, on the next occasion they indulge in a junk-food binge, they are obliged to consume even more high-sugar and high-fat foods to enjoy the same degree of pleasure.

This raises the troubling possibility that dieting – in essence, withdrawal from nice-tasting food – might cause mild mental impairment leading to impoverished decision-making and weaker self-control. It may even be that such impairment, rather than simply a longing for the enjoyment once provided by the now prohibited foodstuffs, is what makes it so hard for people to stay on a diet. This suggests that an arbitrary rule to 'simply' eat less may inadvertently increase the attraction of the forbidden foods and make it more likely they will pile on the pounds. Just as junkies must constantly 'up' their drug doses to experience the same high, so too do junk-food junkies need to consume ever larger and more frequent portions of their favourites once the reward centres in their brain become satiated.

So far in this chapter I have focused on the effects of fat and sugar on brain chemistry and explained why certain foods are so hard to resist. Yet there are other factors involved in stimulating the impulse to overeat. These have nothing to do with how the food tastes and everything to do with the way it is marketed.

## Manufacturing the impulse to overeat

We are worldly wise enough to know that it is better to either avoid, or at least restrict our indulgence in, many activities – such as smoking, drinking and sexual behaviour – which while highly pleasurable have serious health consequences. So why is it that so many of us find it so hard to resist foods high in sugars and fats? Why do we continue to binge on junk food in the knowledge that it adds pounds to our weight and takes years off our life? The answer is that not only do they taste delicious, they are also readily available, convenient and significantly cheaper than healthier unprocessed food. For around a pound sterling in the UK or a dollar in the US, it is possible to purchase over 1,000 calories in the form of biscuits, chocolate or potato crisps. For the same sum you can, typically, obtain less than 300 calories from a bunch of carrots or a bag of apples.[26] According to Dr Gene-Jack Wang, there are many parallels between modern food manufacture and the way cocaine has been produced down the centuries. Over long periods of time people learned how to purify and manipulate the coca leaves so as to achieve a stronger, faster and ever more addictive fix. Food manufacturers have achieved the same results far faster and much more efficiently. 'Our ancestors ate whole grains, but we're eating white bread,' he comments. 'American Indians ate corn; we eat corn syrup.' Corn syrup, used to sweeten almost everything from soft drinks to breakfast cereals, is a staple constituent of highly processed food. This despite the fact it has been linked to an increase in type 2 diabetes, which now afflicts more than one in twelve Americans. Furthermore, meat from cows raised on corn rather than grass is higher in calories and contains more omega-6 and fewer omega-3 fatty acids, a ratio linked to heart disease.[27]

The ways in which high-fat and high-sugar foods are advertised and marketed also play a powerful role in triggering the impulse to eat not wisely but too well. One of the most successful campaigns of recent years was McDonald's tie-in with the massively popular 3D movie *Avatar*. This featured 'the most extensive deployment' of augmented reality ever used in a marketing effort and perfectly illustrates the way entertainment and advertising are becoming fused with the sale, purchase and consumption of food. The campaign, launched on 10 December 2009, included the promotion of *Avatar* character toys, TV commercials and a 'social media outreach', which involved global media webcasts featuring the film's director, James Cameron, together with McDonald's Global Chief Marketing Officer Mary Dillon and USA Chief Marketing Officer Neil Golden. Conducted in more than 40 countries and across 20 languages it involved 25,000 restaurants serving some 45 million customers a day.[28] The goal was to ensure 'deeper engagement with the brand through immersive online experiences' which would ultimately lead to 'repeat purchases of iconic products, like the Big Mac'. In their press release, McDonald's announced: 'Our Happy Meal program specifically focuses on the spirit of adventure and imagination with highly innovative toys, so no matter how old you are, the McDonald's *Avatar* global experience will immerse you in an extraordinary technology-driven journey that's sure to "Thrill the Senses".' Buying Big Macs enabled consumers to access online *Avatar* experiences and each Big Mac featured one of eight '*Avatar* Thrill Cards'. When placed in front of a webcam or mobile phone, the thrill card activated McDonald's branded 'McDivision' software enabling users to enter the virtual Pandora-themed online worlds (which included the McDonald's logo) and play an interactive Pandora Quest game, designed as an 'interactive prequel' to the film. This offered users an 'augmented reality experience', enabling them to 'visit' the *Avatar* rainforest and engage in 'intense action play, simulating the closest thing possible to becoming an avatar in the film itself' and offering a 'subjective and personalised immersive experience'. Followers on Twitter were able to decode a series of daily word scrambles that related to McDonald's and/ or the movie. The winner received an *Avatar* Daily Prize Pack and cinema tickets, the top prize being a private screening with the film's producer. The *Avatar* campaign proved a massive commercial success both for the movie and McDonald's, reaching more than three million customers in the first three weeks. Digital engagement exceeded 10 minutes per customer with an 18 per cent increase in Big Mac sales being reported from Europe, Brazil ('best ever December'), Singapore and Taiwan.

As this example shows, digital marketing offers a powerful and effective way not only of promoting commercial messages to young, impressionable, computer-savvy consumers but of integrating those messages into the very fabric of their lives. By personalising the offer, by making consumption seem not only fun and cool, but an integral part of their self-image. It serves to establish an unconscious association between food cues – like the Golden Arches – and high levels of reward. Environments associated with fast food also come to trigger cravings for high-fat and high-sugar foods. Most fast-food restaurant customers, for example, find it very hard to resist ordering French fries, and possibly a sugary drink, with their burgers. When it comes to making poor food choices on an impulse, the impulse to overeat – massively stimulated by the way food is now manufactured and marketed – all too often results in clogged arteries, type 2 diabetes, sky-high blood pressure and significantly shortened lives.

## Ten practical ways to beat the overeating impulse

There are a number of simple, practical steps that we can all easily follow to fight back against the commercial pressures to overeat.

**1.** Eat from smaller plates and bowls. Because we eat with our eyes as much as with our mouth, using smaller plates and bowls is a quick and easy way to reduce the number of calories consumed at each meal.[29] In one study, when participants were asked to estimate the number of calories in a medium-sized hamburger served on a small plate, they reported that it contained some 18 per cent more calories than when served on a large plate. The same applies to desserts served in bowls. In a study I conducted in London as part of a Channel 4 television documentary entitled *Secret Eaters* we found that volunteers who served themselves ice cream into a large bowl took 44 per cent more than those with a smaller bowl.

**2.** When drinking anything but water, use a tall narrow glass rather than a short wide one. Research[30] shows that even experienced barmen will pour more drink into a short wide glass than a tall narrow one when asked to pour an exact measure without the use of a measuring device. This is known as the Horizontal–Vertical illusion. The eye is fooled by the height of the liquid in the glass and ignores the width. In a study I conducted, those given a wide tumbler poured almost twice as much as those with a narrow glass (397 ml vs. 796 ml).

**3.** Eating with chopsticks, rather than a knife and fork, obliges you to take smaller mouthfuls and eat more slowly. Eating more slowly allows time for the digestion process to work effectively. It also gives you a better chance of recognising when you have eaten sufficient food. Around 20 minutes is required for signals from the stomach to reach the brain and let you know that it is full. Research shows that the average meal is consumed in around 12 minutes.

**4.** Put ice in your drink. Because the body has to use energy to heat up the beverage, around 1 calorie per ounce of fluid is consumed. If you drink the recommended eight 8-ounce glasses of water a day with ice cubes you will burn up between 60 and 70 extra calories each day.[31]

**5.** We eat significantly more when dining in company than eating alone. For *Secret Eaters*, I asked diners to have lunch alone, as a couple or at a table of six friends. The group consumed some 600 more calories than the solitary diner and over 80 more than the couple. What happens is that we tend to pace ourselves on the fastest eater in a group. Also, distracted by conversation and banter we fail to notice how much we are eating and are more susceptible to offers of second helpings. Be aware of this risk the next time you eat with a group of friends or colleagues. Refuse second helpings. Pace yourself to the slowest eater at the table. Try to be the last person to start eating.

**6.** When dining out in an upmarket restaurant be aware of the effects of soft lights and classical music. Both encourage you to linger longer and so eat or drink considerably more than you realise.

**7.** Get a good night's sleep. Research from Sweden shows that a specific brain region that contributes to a person's appetite sensation is more activated in response to food images after one night of sleep loss than after one night of normal sleep. Using functional Magnetic Resonance Imaging (fMRI) to measure changes in blood flow to different regions of the brain, Christian Benedict of the Department of Neuroscience at Uppsala University found that after losing a night's sleep 'males showed a high level of activation in an area of the brain that is involved in a desire to eat. Bearing in mind that insufficient sleep is a growing problem in modern society, our results may explain why poor sleep habits can affect people's risk to gain weight in the long run. It may therefore be important to sleep about eight hours every night to maintain a stable and healthy body weight.'[32]

**8.** Avoid shopping for food when hungry. In one study, I compared the purchases of a group of volunteers who had been starved for seven hours prior to going shopping with a second group who had been provided with healthy snacks during the morning. The hungry shoppers bought foods higher in calories (2,840 vs. 715), in fat (141 g vs. 28 g) and sugar (118 g vs. 48 g) than those who had taken the edge off their appetites.

**9.** If you have sweets or chocolates in the office, place them in opaque containers rather than clear ones and at a distance from, rather than directly on, your desk. When we tested the difference in a busy London office the results were startling. Workers provided with sweets in a clear glass bowl on their desk consumed 500 per cent more than when the treats were hidden from direct sight and required slightly greater effort to reach them.

**10.** When snacking on popcorn in the cinema, use your other hand. That is, if you normally hold the popcorn container in your right hand and pick with your left, then swap around. By making this small change in your eating habits you force yourself to think more about what you are doing and are, therefore, likely to eat less. During a study to test the effects of this tactic in a London cinema, I provided half the audience with an oven glove worn on the normal popcorn picking hand. Their consumption fell by a quarter compared to their non-handicapped companions.[33]

In the next chapter, I shall be examining the ways in which retailers seek to encourage the purchase of high-fat and high-sugar foods through the design, display and layout of their stores.

# CHAPTER 10

## The Buying Impulse – The How and Why of What We Buy

'Twenty million seconds. That is the time all consumers
collectively spend in a typical supermarket every week.
Each one of those seconds is an opportunity to sell. That
is 20 million opportunities a week to sell something.'
Herb Sorensen, *Inside the Mind of the Shopper* [1]

While making a TV documentary about supermarkets, I asked
shoppers leaving the store whether or not they had bought
anything on impulse. All but one agreed they had. Only a middle-
aged man denied having done so. 'I make a list and stick to it,'
he told me rather smugly. Then added, 'I'm a supermarket
manager and I know all the tricks!' For many shoppers his answer
will confirm their suspicion that supermarkets are in the business of cunningly persuading them to impulse-buy things they
don't really need and may not easily be able to afford.

To be sure, supermarkets have elevated the art and science of
persuasion to previously unimagined heights and made impulse
buys – 'splurchases' as they are called in the trade – very big business indeed. It has been estimated that in the UK and USA alone,
shoppers currently spend some £24 billion a year on impulse buys.
In Britain these account for between 45 and 100 per cent of retail
turnover while in the United States approximately 62 per cent of
supermarket sales and 80 per cent of sales of luxury goods are
made up of impulse purchases. Surveys have shown that nine out

of ten consumers impulse-buy at least one item per shopping trip, with more than half admitting to as many as six. This is estimated to amount to a lifetime spend of around £50,000 for each individual.

## The psychology of the impulse buy

What motivates consumers to impulse-buy has been the subject of research and debate among retailers and psychologists for over 60 years. One of the earliest papers on the topic was published in 1951 by William Applebaum of Stop & Shop Inc.[2] He suggested that impulse buying might be due to the shopper's exposure to sales promotion and advertising materials. Later researchers expanded on these ideas, proposing a variety of reasons for making such purchases. In 1962 Hawkins Stern, an industrial economist at the Stanford Research Institute in southern California, identified four main types of impulse buy.

1. *Pure impulse buying* is a novelty or escape purchase which breaks a normal buying pattern.
2. *Reminder impulse buying* occurs when a shopper sees an item or recalls an advertisement or other information and remembers that the stock at home is low or exhausted.
3. *Suggestion impulse buying* is triggered when a shopper sees a product for the first time and visualises a need for it.
4. *Planned impulse buying* takes place when the shopper makes specific purchase decisions on the basis of price specials, coupon offers and so forth.[3]

As we shall see in a moment, supermarkets devote a great deal of time, energy and money to identifying the most effective ways of pressing each of these four 'buy buttons'. Some 20 years later

Dennis Rook, a research associate with DDB Needham Worldwide, suggested that impulse buying – which he described as 'pervasive . . . extraordinary and exciting' – occurs 'when a consumer experiences a sudden, often persistent urge to buy something immediately. The impulse to buy is . . . prone to occur with diminished regard for its consequences'[4] In the past, many psychologists took a negative view of impulse buying, perhaps due to the lack of perceived control with which it is associated. More recent research suggests, however, that shoppers do not, for the most part, consider it a mistake to make spur-of-the-moment purchasing decisions.[5] Only around one in five express any regret for their impulse buys, with four out of ten claiming to feel good about them.[6] In this chapter, I will be exploring the extent to which consumers' suspicions that some retailers seek to manipulate their minds and undermine their self-control, so as to encourage impulse purchases, are justified. My focus will be mainly on the techniques used by supermarkets[7], not because they are uniquely manipulative but because they are uniquely successful. They are also where most shopping is now done and where most consumers spend the greatest amount of time shopping. In the near future they may also be the only places where much of our weekly shopping *can* be done. As Joanna Blythman wryly points out in her book *Shopped*: 'If supermarkets get their way, everything we buy, whether it is from a superstore, a smaller city store, a corner shop or a petrol-station forecourt, will be sold by a supermarket chain. There will be small shops, but not as we know them.'[8]

## Impulse buys and the rise of supermarkets

Ironically, in view of the extent to which out-of-town stores have destroyed the traditional high street, supermarkets were originally established not by faceless corporations but by independent retailers seeking to challenge the growing power and dominance of major retail chains. They were essentially an expansion of an earlier retailing innovation, that of self-service shopping. The world's first supermarket[9] was opened in a disused garage in Jamaica, Long Island, New York on 4 August 1930, by a flamboyant grocer named Michael J. Cullen. Inspired by the movie about the giant ape King Kong, Cullen named his store *King Kullen* and boasted it would be the 'World's Greatest Price Wrecker'.[10] Among his innovations was a separate food department selling large quantities of food at discount prices and a parking area. By the time of his death, six years later, his King Kullen stores had grown to 17. Yet, despite their obvious popularity and financial success the established grocery chains ignored them. When they could be bothered to comment on these upstarts at all, their tone was pompously dismissive. They described their rivals as 'cheapy', 'horse and buggy', 'cracker barrel storekeeping' and 'unethical opportunists'. A senior executive at one major chain remarked that he found it hard to believe 'people will drive for miles to shop for foods and sacrifice the personal service chains have perfected and to which Mrs Consumer is accustomed'. In 1936, speakers at the National Wholesale Grocers convention and the New Jersey Retail Grocers Association insisted traditional retailers had nothing to fear. They assured their audience that 'price buying' appealed to only a small number of shoppers and that the limited size of the market would soon stop supermarkets dead in their tracks. Most consumers, they proclaimed, wanted personal service and were not interested in discounted prices.

In the years that followed, however, more and more retail chains, recognising the strength of the coming storm, moved into supermarkets themselves. Even though this involved the loss of their huge investment in corner stores and established methods of distribution and merchandising. 'Many people find it hard to realise there ever was a thriving establishment known as "the corner grocery store". The supermarket has taken over with a powerful effectiveness,' wrote Theodore Levitt.[11] 'Yet the big food chains of the 1930s narrowly escaped being completely wiped out by the aggressive expansion of independent supermarkets . . . The companies with "the courage of their convictions" resolutely stuck to the corner store philosophy . . . They kept their pride but lost their shirts.'

## Inside the world's most powerful selling machine

To regard a modern supermarket as a shop is to miss the point. These often massive cathedrals of consumerism are far more than steroid-pumped versions of the familiar high-street grocer's shop. Rather, they are meticulously designed and engineered selling machines whose sole purpose is to supply consumers with their necessities and do everything possible to stimulate their desires. Just by walking into a supermarket, we become participants in what Jeff Chester, director of the US-based Center for Digital Democracy, has called a 'marketing technological "arms race"'. Everything we see, hear, smell, taste or touch is the result of millions of pounds worth of research, design and planning. While shoppers often blame their lack of willpower, the truth is that their impulse buys are due less to irrational behaviour than to the sophistication of modern marketing, advertising and retailing strategies. Pass through the supermarket entrance and you enter the kingdom of the impulse buy.

Let's visit a typical large supermarket. Studies have shown that we will spend around 29 minutes inside – seven minutes less than we would have done a decade ago – mostly walking from one display to the next and one food category to another. During this time we will travel about half a kilometre if making a small basket shop and around a hundred metres further when pushing a trolley.[12] Few if any attempts are made to sell to us immediately beyond the entrance. There may be a photo-booth, some charity collection boxes and perhaps a ride-on car or horse for toddlers. Extending some 20 feet into the store, this is what retailers term the 'decompression zone'. Its purpose is to ensure that shoppers adjust, physically and psychologically, to their new surroundings. The pace at which we walk down a street or across a car park is a good deal faster than the 'grazing' speed at which supermarkets managers want us to move through their stores. The 'decompression zone' not only slows us down but also helps us adapt to changes in light levels, temperature and humidity. Once this adaptation has been achieved we will be in the optimum frame of mind to impulse-buy.

## The law of the aisles

The first products we typically see on entering a supermarket are fresh fruit and vegetables. From the shopper's point of view their location, close to the entrance, makes little sense. Vegetables, after all, tend to be heavy and bulky. Not items to be carried around for the remainder of the shopping trip. In addition to their weight, many fruits are also easily bruised. It would seem to make more sense to place them closer to the exit so that they could be purchased at the end of the shopping trip rather than at the start. But this would be to mistake consumer psychology for retailer psychology. They are positioned at the front of the shop for two excellent reasons. First because they make both the store and the products themselves look fresh and attractive. Fruit and vegetables always look better in natural light, just as meat and fish look tired in anything but a clean, white light. Furthermore, they subtly convey a sense of freshness and naturalness, evoking images not of vast and soulless food-processing plants but of open green fields and cloudless blue skies. The second and chief reason is to make the shopper more likely to grab a trolley than reach for a basket. Because a trolley holds more it encourages shoppers to buy more. By making the task of moving purchases around far easier it encourages shoppers to impulse-buy more.

While fresh fruit and vegetables are placed at the front, other groceries are displayed much deeper into the store. This encourages shoppers to peruse food items with a higher profit margin, which they might be tempted to impulse-buy, on their way to find the goods they actually want. The shelves, rather than running the length of an aisle, are usually divided into several short sections. This enables the retail geographers, responsible for the layout, to increase the number of 'aisle ends', or 'end-caps' as they are sometimes called. 'The aisle ends are the monthly engines of the business and the promotional calendar is driven by their performance,' says Karl McKeever, Brand Director for Visual Thinking. 'They are in high traffic positions, so lots of people will see them.'[13]

Another factor retailers must take into account when stacking shelves is known as the invariant right. From early infancy, children will approach and reach out to objects using their right hand more frequently than their left. A possible explanation for this is that the left hemisphere of the brain controls the right side of the body and this hemisphere is associated with an approach response and positive emotions. The right hemisphere, by contrast, controls the left hand, and this is associated with an avoidance response and more negative emotions. If you are right-handed and pick something off the shelf on an impulse, then not only will it most likely be with your right hand but you will also value your purchase more highly than if you had used your left. Shoppers tend to stay on the right as they stroll down both supermarket aisles and high-street pavements. It is for this reason that, in a well-designed airport, travellers drifting toward the departure gate will find the fast-food restaurants on their left and the gift shops on their right. Both are in positions most likely to encourage impulse purchases: while people may be prepared to cross a lane of pedestrian traffic when hungry they will rarely do so to buy a magazine or souvenir.

Research has shown that when we walk down an aisle we look mainly at shelves not at eye level but slightly lower. This is because, when walking past the shelves, a shopper's gaze is directed between 15 and 30 degrees downward, mainly as a result of the weight and shape of the head and how it is supported by the spine.[14] This, then, is the prime position for high-margin impulse purchases. Lower priced, lower profit products go either down to floor level or upwards so as to be out of reach to all but the tallest or most agile shopper.

Going deeper into the store we come across products termed 'destination goods' or 'known value items' (KVIs). KVIs include such core grocery lines and weekly staples as milk, bread and baked beans. Supermarkets call these 'traffic generators' because they have to be purchased frequently and are the most price-sensitive. While the average shopper's awareness of prices tends to be limited, most do know the price of these regularly bought items and so are able to make comparisons between rival stores. By ensuring the price of benchmarked KVIs is kept artificially low, by selling at or even below cost, supermarkets are able to present *all* their products as great value for money. This, as any careful shopper knows, tends to be far from the truth. Bargain prices on KVI's help convince shoppers the store is 'on their side' in keeping down the cost of living. As Tesco likes to boast, 'Every little helps.'

## Impulse buys and moments of truth

Retailers and brand managers talk a lot about the 'moment of truth'. This is not a philosophical notion, but the point when people standing in the aisle decide what to buy and reach to get it. The strategy is to exploit psychological tactics to 'nudge' the customer's hand in the direction of a particular product. If presented with two equally attractive items customers may be unable to decide between them and end up buying neither. To avoid this undesirable outcome some supermarkets introduce a third, 'decoy' item, deliberately made in some way less attractive than the other two. Being able to quickly and confidently reject this option encourages shoppers to impulse-buy one of the other two. On counters displaying clothes, high-margin items such as expensive and exotic foods, enticing gifts, must-have gadgets, perfumes and grooming products, these products are frequently laid out in such a way as to exploit the power of what is called the triangular balance. 'The triangular balance is based on the fact that your eye will always go straight to the centre of a picture,' says Karl McKeever. 'Here, they put the biggest, tallest products with the highest profit margin in the centre of each shelf and arrange the other sizes around them to make it look attractive. When you look at the triangle on the shelf, your eye goes straight to the middle and the most expensive box. It is used everywhere and it's very effective.'[15] Having admired and desired the best that money can buy, it's hard for consumers to settle for anything less. We impulse-buy luxury items as a reward, either for someone dear to us or, as frequently, for ourselves. The more expensive that treat the better we feel about ourselves. Almost all of us think we are 'worth it' one way or another and will never settle for 'second best' even if this is all we can really afford.

'A man's Self,' wrote the 19th-century psychologist William

James, 'is the sum total of all that he can call his, not only his body and his psychic powers, but his clothes and his house, his wife and children, his ancestors and friends, his reputation and works, his land, and yacht and bank account. All these things give the same emotions. If they wax and prosper, he feels triumphant; if they dwindle and die, he feels cast down.'[16]

## Touch me, feel me, impulse-buy me

By encouraging customers to touch, handle, play with or consume the products the probability of impulse purchases is significantly increased. Free samples of cheese, crisps, snacks, chocolates or fizzy drinks are all used within stores to boost spur-of-the-moment buying. Some supermarkets will allow shoppers to return non-food items if dissatisfied on returning home. While this enhances confidence in the impulse buy and makes the store more shopper-friendly, the actual cost of dealing with returns tends to be relatively modest. For one thing, lethargy or forgetfulness means that many shoppers never bother to return goods, especially low-value items. They just stick them in a cupboard and forget them. A second and more important reason is that having taken possession of an item we regard it as having enhanced emotional and financial value. Known as the 'endowment effect' – or 'divestiture aversion' – it is one of the most robust findings in the science of behavioural economics. It's the reason why tossing out that pair of comfortable but tattered slippers or throwing away a favourite pair of thread-bare jeans can prove so surprisingly hard. It also helps explain why, long after childhood's end, grown adults hang on to their treasured doll or teddy bear. The most widely accepted explanation for this effect is loss aversion. That is to say, we experience greater pain when losing something than we derive pleasure from owning it in the first place. Retailers know this from practical experience, with some salesmen dubbing the technique 'taking the puppy home'. They know that if a shopper can somehow be made to feel ownership of an item, even on a temporary basis, they are far more likely to want to keep it.

In fashion areas of stores, tables are often positioned next to clothes racks to allow customers to handle the items. Younger shoppers especially are drawn more towards scuffed-up displays of clothes because this suggests they are popular. When the piles of garments appear overly neat, by contrast, it sends a message that no one else is interested in buying them. As we shall see in the next chapter, interest from others significantly increases the excitement and arousal that triggers impulsive behaviours. This applies whether the action is buying a product for which there is a strong demand (think of the panic buying during annual sales), rioting, looting or even taking one's own life. If no one is doing it then few are interested. If many are doing it then many others will want to do it.

As with clothes, so too with other items. Disordered piles and mounds of products are more likely to attract buyers than neatly ordered displays. There is a natural reluctance on the part of most shoppers to be the first to interfere with the symmetry of perfectly stacked tins of food or ordered lines of fruit in an elegant display.[17] Most products are handled many times by browsers before being purchased. The average lipstick, for example, is examined six to eight times before it leaves the store and a greeting card 25 times. The more often and the more intensely people can be encouraged to interact with a product, play the latest video game, try on the latest fashion, fly the toy helicopter around the store, sample a chocolate or spray on a perfume the more likely it is that feelings of ownership will kick in and the endowment effect help close the deal.

# Junk food is cheap food

Immediately after the decompression zone of many supermarkets you will often find high-fat, high-sugar impulse buys with items such as crisps, chocolates, biscuits and other sweet temptations placed on easy-to-reach shelves or even in open bins where the customer only needs to dip in a hand to satisfy a craving. In studies of supermarkets in different neighbourhoods, I have found this easy – one might almost say irresistible – way of displaying junk foods more likely to be present in poorer, working-class areas than in wealthier and more middle-class ones. By locating these displays in an area where shoppers will pass them on both entering and leaving the store, the temptation to combine impulse eating with impulse buying is doubled. Furthermore, these products are often a very low-cost source of energy. One brand of biscuits, for example, on sale at £3 a tin in the North of England will cost you £16 when more stylishly packaged and sold in a West End London store.

The reason why many high-fat, high-sugar foods are relatively cheap is because the materials used in their manufacture, such as corn syrup, air and water – all of which are used extensively – cost little. Between 1985 and 2010 in the USA, for example, the price of fresh fruit and vegetables rose by nearly 40 per cent. Over the same period, the price of beverages sweetened with corn syrup, which is high in fructose, fell by a quarter. According to the National Soft Drink Association, soft drink consumption in the USA has doubled for females and tripled for males. Among men aged between 12 and 29 average consumption is now 160 gallons a year. Every last drop of it sky-high in calories and rock-bottom in nutrition. By comparison, UK consumption was a far more modest but still fattening 21 gallons, with Germans showing even more restraint at 16 gallons and Japan currently supping the lowest amount of sugary drinks at just 5 gallons.

## Eliminating barriers to impulse buying

In addition to increasing the time shoppers spend browsing the aisles – so called 'dwell-time' – retailers are skilled at detecting and eliminating barriers to buying. One of these, first identified by the American retail anthropologist Paco Underhill, is the 'butt-brush' factor. 'Women especially, although it is also true of men, to a lesser extent – don't like being brushed or touched from behind,' he reports.[18] 'They'll even move away from merchandise they're interested in to avoid it.' Research for a major bookstore chain identified the butt-brush factor and the need for women to watch children as major obstacles in converting browsers to impulse buyers. The chain was advised to adopt a store layout with angled instead of straight aisles to create browsing 'nooks' for displays at the end of each aisle.

While they do not, as a rule, set out to cheat their customers, supermarkets certainly do all they can to put temptation in their way. To control what they see, hear, taste, touch and feel in such a way as to direct them to those products they are most eager to sell and the shopper is most likely to purchase. In short, to do everything and anything they can to encourage the impulse buy.

# CHAPTER 11

# The Imitation Impulse – 'A Beautiful Place to Die'

'Each to each a looking glass reflects the other that doth pass,' Charles Horton Cooley, *On Self and Social Organization*[1]

Yawn in a crowded room and others will yawn too. Yawning is infectious. So is laughter. Which is why a studio audience, or at least a recorded 'laughter track', is used on TV comedies. It encourages viewers to think what they are watching really is funny. Cry and those around you are likely to feel more miserable. Become angry and others may follow your example. Throw a brick through the window of an abandoned building and within hours every pane of glass will have been smashed. Leap to your feet at the end of a performance and, almost at once, the rest of the audience will be giving the performer a standing ovation. That we are all natural copycats is a fact advertisers and marketers have long known and exploited. One of the most effective ways of persuading people that a new product or service is a 'must-have' buy is to demonstrate that a great many others have already bought it. There is a certain critical mass of consumers – author Malcolm Gladwell refers to it as the 'tipping point' – after which commercial success is assured. As the old saying puts it, 'Nothing succeeds like success.'

While anonymous members of the public can trigger a 'copycat' stampede provided their numbers are sufficiently massive, a far faster and more effective commercial strategy is to use a celebrity.

Which is why famous sportsmen, singers and actors, and even those with the transient celebrity of having appeared on a reality TV show, are paid large sums to endorse brand names.[2] The bigger the name, fame and following a celebrity has, the greater the numbers who will copy their behaviour. There is nothing especially surprising about this. Most of what we know we have learned by copying others. Imitation is the most widely used form of learning, especially during childhood, because it saves the time and effort of acquiring new skills through trial and error.

Sadly, imitation also plays a significant role in the impulses that lead to more violent and destructive behaviours, including rioting, vandalism, looting and suicide. Such activities may not, at first glance, seem to involve impulsive behaviour. It can take hours for a peaceful crowd to transform itself into a rampaging mob and turn legitimate protest into an orgy of violence, arson and looting. By the same token, those who take their own lives at 'suicide hot spots' such as Beachy Head in Sussex or the Golden Gate Bridge in San Francisco[3] have to plan their demise with considerable care and determination. They must travel – often for hours or even days – to reach the place where they have chosen to die. One man flew to England from the USA and then travelled the 70 miles down to Sussex to leap from the cliffs at Beachy Head. Before doing so he engaged undertakers, bought a coffin and even arranged the flight number and date for it to be shipped back to the States with him inside.[4] Another walked the length of the beach surveying the cliffs until he found a straight drop from which to jump. Alerted by a member of the public, the police were able to detain the man before he could do so.[5] Potential suicides often write a 'goodbye' note to their loved ones and ensure these are kept safe. One elderly woman who jumped to her death from the cliffs near Beachy Head had first carefully sealed farewell letters in plastic bags before putting them into her pocket. She wanted to ensure they would still be legible should her body end up in the sea.[6] How, then, can such behaviour in any way be described as impulsive?

The same question could, of course, be asked about many sorts of impulsive behaviours. Even falling in love at first sight usually involves planning and organisation in order to meet Mr or Ms Right. So too does buying things or overeating. What, then, is 'impulsive' about any of these behaviours?

In Chapter 2 I presented examples of Necker Cubes, which abruptly change their shape as you look at them. The switch from one to another is rapid, unpredictable and outside our ability to control. An impulse works in a similar manner, flicking our brain from System R to System I thinking in an instant. It occurs in that moment when the eyes of lovers meet, a 'must-have' product is taken from the display shelf or a second helping of Black Forest gateau guiltily accepted. Within this typically very brief time frame the rational brain takes a back seat and our zombie brain takes over. In this chapter I want to examine the factors that contribute to this switching in two dramatic and often tragic circumstances. Riots and suicides.

## The impulse to riot

At around 6 p.m. on Thursday 4 August 2011, a 29-year-old black man named Mark Duggan was shot dead by the police in London in circumstances still not fully explained.[7] On Saturday 6 August, at about 5 p.m., a group which included members of Duggan's family gathered outside the local police station, in peaceful protest. They wanted to speak to a senior police officer about the killing but no one was available. Over the next few hours the crowd grew to some two hundred all demanding the same thing. An explanation for Duggan's death. No senior officer emerged to hear their grievances. At around 8.30, some three and a half hours after the protest had begun, missiles were thrown at parked police patrol cars. Shortly afterwards two more police cars were attacked, set alight and pushed across the road to create a barrier between the crowd and a police line. The offices of a local law firm were set ablaze as was a bookmaker's. Other fires soon followed. Over the course of the next few hours, shops were ransacked and burned out, cars and a bus set on fire, members of the public assaulted and robbed. The riots[8] rapidly spread to other parts of London and to other English cities. 'The speed of the disintegration said everything,' wrote *Financial Times* journalist Gautam Malkani. 'It took less than 48 hours for London to descend from self-styled capital of the world into a circuit of burning dystopian hells . . . many rioters in London and other cities were laughing as they looted. The speed of the destruction was partly a function, then, of their sheer exuberance – the opposite of stereotypical listlessness more commonly known as "chillaxing". Like football hooliganism the violence was recreational – a day out in a Nietzschean theme park.'[9] The causes of the London riots were numerous, complex, still poorly understood and often inadequately and inaccurately reported. As Steve Reicher, Professor of Psychology at the University of St Andrews, and Cliff Stott from the University of Liverpool comment in their book on the riots, *Mad Mobs and Englishmen*: 'For four nights in August, England burned. Or rather – and the distinction is important – for four nights in August the media was full of images of England burning.'[10]

The media has an unquenchable appetite for drama and the rioting provided a superabundance of drama. As a result, the events of those four nights came to be defined by a few iconic images. A 150-year-old furniture store that had survived two world wars being gutted by fire. Shops being casually and methodically looted.

Rioters assaulting and robbing a young student under the pretext of helping him. A woman silhouetted against flames as she leapt from a blazing building into the arms of a fire officer. These images of the riots, say Reicher and Stott, 'stand for everything that went on. They . . . guide our understanding and our discussion of events.'[11] Far less often discussed was the fact that the riots first erupted in the 13th most deprived borough in England, which is also one of the most ethnically mixed. Seven months earlier, spending cuts had led to the closure of eight out of thirteen youth clubs and the removal of financial support for other services such as after-school clubs and employment support. As I had found in Belfast 40 years earlier, the seeds of discontent and social alienation were to be found in poverty, a lack of opportunity for young people (especially blacks and Asians), envy, greed, gang culture, opportunism, criminality and an absence of parental role models. 'We'd have gone ourselves if we'd been a bit younger,' the mother of one teenage rioter told a reporter.[12] So a case can be made against both the police handling of the initial protests and the way the media reported, and in some instances misreported, the riots.

Here I want to focus only on a few individual cases of impulse-driven crime. The looting and stealing by previously honest citizens who, individually, would never have broken the law but who, as part of a crowd, experienced an overwhelming desire do so. People like the 19-year-old millionaire's daughter charged with being a getaway driver for looters. Well educated and waiting to go to university, she was described as a 'polite, lovely and popular' teenager. Then there was a 'talented sportswoman' who was said to have picked up masonry and hurled it through a shop window; a ballerina, aged 17, alleged to have been part of a mob that stole £90,000 worth of goods from a hi-fi shop. There was a second-year law student remanded in custody after being accused of being part of a gang which ransacked cafés and restaurants; a 22-year-old student of accountancy and finance, a civil engineering student, an estate agent and a would-be chef. Children too were equally caught up in the violence. An 11-year-old girl was arrested after hurling stones at the windows of two stores. A boy of the same age stole a £50 waste bin. Natasha Reid, a 24-year-old university graduate, was preparing for a career as a social worker when she stole a flat-screen TV from a looted store. Her mother, Pamela, said: 'She didn't want a TV. She doesn't even know why she took it. She doesn't need a telly.' Natasha herself was reported as saying: 'I don't get it. Why did I do it?' In the cold light of the following day very few of those accused of taking part could explain what had led them to break the law. Many, like Lorraine McGrane, 19, who was spotted walking away from the rioting with a flat-screen TV, could only say they knew what they had done was wrong 'but everyone else was doing it'.

## Suicide by imitation

Sadly, some of those intent on taking their own lives seem to adopt the same strategy. They do it at a particular location or using a specific method[13] because other people have done it before them and these deaths have been well publicised. Especially in the printed media. 'Unlike televised suicide stories, newspaper suicide stories can be saved, reread, displayed on one's wall or mirror, and studied,' comments Dr Steven Stack of the Department of Criminal Justice at Wayne State University. 'Television based stories on suicide typically last less than 20 seconds and can be quickly forgotten or even unnoticed. Detailed studies of suicides occurring during media coverage of suicide have often found copies of suicide news stories near the body of the victim.'[14]

The risk of copycat suicides has long been recognised. In 1774 Goethe's novel *The Sorrows of Young Werther*, in which the hero commits suicide as a result of a failed love affair, was banned in many parts of Europe after it had led to a rash of imitative suicides, particularly among young people. The suicide of a celebrity almost always leads to a series of copycat suicides. While there is only a small increase, of around 3 per cent, over the month following the death of an unknown, suicides increase by as much as 14 per cent when a celebrity takes his or her own life. After film star Marilyn Monroe's reported suicide in August 1962, for example, there were 300 additional suicides – an increase of 12 per cent. When *Final Exit*, a guide to suicide for terminally ill people that advocates death by asphyxiation, was published in 1992, the number of suicides by asphyxiation in New York City rose from 8 to 33, an increase of around 400 per cent. A copy of *Final Exit* was found at more than a quarter of these suicides.

## Peer pressure and the suicide impulse

Mass suicides may involve as few as two people, in a 'suicide pact', or larger numbers of people who are persuaded or compelled to kill themselves by a charismatic leader. Examples include the deaths of members of the Reverend 'Jim' Jones's People's Temple in Guyana in 1978. More than 900 cult members, including 200 children, died after queuing up to drink Flavor Aid laced with cyanide. In 1997, 39 members of the San Diego-based UFO cult, Heaven's Gate, killed themselves after being convinced by their leader, Marshall Applewhite, that they would wake up aboard a flying saucer that was following Comet Hale-Bopp.

The internet too is applying peer pressure, especially on young people, via websites that either support or strongly advocate suicide. Indeed, the rapid growth of such sites – a Google search today yields some 243,000 hits compared with around 100,000 ten years ago – has been cited as a possible explanation for the fact that suicide is now the leading cause of death among teenagers and adults under the age of 35.[15] 'Suicidal adolescent visitors risk losing their doubts and fears about committing suicide,' explains leading expert Katja Becker, from the Central Institute of Mental Health in Mannheim, Germany. 'Risk factors include peer pressure to commit suicide and appointments for joint suicides. Furthermore, some chat rooms celebrate chatters who committed suicide.'[16]

By taking their own lives at a famous 'suicide' location or adopting the same methods as an admired celebrity, confused and depressed individuals may be finding comfort in the fact they are not alone or unusual in their urge for self-destruction. Occasionally the role played by others in encouraging suicides can be even more direct. With the spread of mobile phones, people who decide to kill themselves in public are likely to end up on YouTube, Facebook and Twitter. The crowd that quickly gathers whenever someone threatens suicide can sometimes play a far more active and malicious role than merely recording the event on their mobile phone.

One Saturday afternoon in January 2008, 17-year-old Shaun Dykes climbed to the roof of Westfield shopping centre in Derby city centre and stood on the edge preparing to jump. Detective Inspector Barry Thacker, a police negotiator, arrived and tried to talk him down. A crowd of about 300 gathered on the pavement below. Many started filming the event on their mobile phones. One yelled 'How far can you bounce?' while others urged him to 'get on with it'. Shaun was engaging with us,' Barry Thacker told the inquest. 'I had been sat with my arms out towards him. He bent down to reach down to my hand when there was a shout of, "You're wasting taxpayers' money". He stood up, said, "It's gone too far" and started counting down. Then he said, "No", closed his eyes and threw himself off.' He hit the pavement 60 feet below and died instantly.[17]

## When second thoughts come too late

Tragically, within seconds of succumbing to the suicidal impulse, an individual can bitterly regret their decision, as an incident recounted to me by Sussex coastguard Don Ellis illustrates. Don, a stocky, bearded man in his early sixties, is stationed close to Beachy Head, on England's south coast. At 530 feet, the UK's highest chalk sea-cliff, it attracts more than a million visitors a year. Most come to stroll along the cliff paths, picnic on the soft turf and enjoy the breathtaking views. Others, however, have far sadder reasons for making the journey. The headland is, in the words of novelist Louis de Bernières, so beautiful that it 'openly invites you to die'.[18] Each year around 50 people do just that.

Don was on his way to the Coastguard Station when he spotted a middle-aged man walking close to the cliff edge. The man was smartly dressed and he was carrying a brightly coloured umbrella. Highly experienced in the ways of potential suicides – he has had to deal with well over a thousand during his two decades in the service – Don moved slowly and calmly towards him. The reason why he kept his movements slow and casual was to avoid startling the man into jumping. When they were some ten feet apart, he remarked cheerfully, 'Good morning. Not a very nice day is it?'[19] The man smiled but said nothing. He carefully rolled up his umbrella and stuck it into the earth. Then, still without uttering a word, he stepped over the edge. As he tumbled to his death on the rocks 500 feet below Don heard him scream: 'Help me! Help me!' 'At that point,' Don remarks sadly, 'there was nothing anyone could have done.'

## Monkey see – monkey do

During the early 1990s, Italian researchers made a remarkable and
entirely unexpected discovery. A discovery which, according to the
noted neuroscientist Vilayanur Ramachandran, represented 'the
single most important "unreported" (or at least, unpublicized)
story of the decade.'[20] While studying brain functions, the scientists
had implanted electrodes in macaque monkeys. One day, a
researcher felt hungry and decided to have a snack. As he reached
for his food, he noticed that neurons in the monkey's pre-motor
cortex had started to fire in the same area that became active when
the animals started reaching for their own food. But why was it
happening when the macaque was merely sitting still and watching
him?

The answer, the researchers subsequently found, lay in the exist-
ence of specialised neurons which fire in response not to the actions
of the individual but to the activity of others. They called them
'mirror neurons' and, 20 years later, few scientists doubt the importance of this discovery. 'Mirror neurons will do for psychology
what DNA did for biology,' predicted Ramachandran, 'they will
provide a unifying framework and help explain a host of mental
abilities that have hitherto remained mysterious and inaccessible
to experiments.'[21] In the decades following that serendipitous
discovery, this special class of brain cells led to a dramatic shift in
the way psychologists view the brain, especially in its social aspects.
Prior to their identification there was a general belief in the scientific community that we interpret and predict the behaviour of
others by using rational, System R thinking. They now understood
that social understanding and empathy were likely to arise automatically as mirror neurons 'simulate' not only the actions of
others but also their intentions and emotions.[22] When we see
someone smile, for instance, our mirror neurons for smiling also

fire, creating in our mind the emotions associated with the smile. There is no need for us to think about the other person's intentions behind the smile – the experience is immediate and effortless.

Building on that initial research with monkeys, neuroscientists now use brain imaging to investigate mirror neurons in humans.[23] Among the key findings from these studies is the fact they allow us to determine the intention behind the action. When someone picks up a cup of tea to enjoy a drink, for example, our mirror neurons become more active than when the person picks up the same cup while clearing the table. It appears that the system enables us to decode facial expressions. The same region in the brain fires whether we wrinkle up our nose in disgust or merely observe someone doing the same. This occurs even when the movements involved are so slight and subtle as to be outside conscious awareness. The 'microgapes' associated with expressions of disgust, as described in Chapter 6, trigger the same response in anyone watching.

Studies have suggested that people with autism, a condition characterised by an inability to empathise with others, may have a dysfunctional mirror neuron system. The more severe the autism, the lower the activity recorded in their mirror neuron system. This may, in part, be due to their inability to look at other people's facial expressions, so-called 'gaze aversion'. In a study conducted in my own laboratory, autistic participants were shown a specially recorded drama. The short sequence depicted a man who is having an affair with his wife's best friend. Although non-autistic members of the audience twigged what was going on almost immediately, none of those with autism had the slightest idea. When we examined where they had been looking, using eye-scanning technology, it was clear that while the non-autistic viewers had concentrated on the actors' expressions the gaze of the autistics had ranged all over the scene without once being directed to any of the three faces.

Watching a riot, whether in person or on a screen, will therefore trigger mirror neurons and the emotions of excitement, anger, fear or rage associated with the actions being observed. The responses triggered by mirror neurons are reinforced by the presence of others to such an extent that the impulse to act in the same way may prove overwhelming.

## The power of others

In his influential book *The Crowd*, the 19th-century French writer Gustave Le Bon remarked: 'Whoever be the individuals that compose it, however like or unlike be their mode of life, their occupations, their character, or their intelligence, the fact that they have been transformed into a crowd puts them in possession of a sort of collective mind which makes them feel, think, and act in a manner quite different from that in which each individual of them would feel, think, and act were he in a state of isolation.'[24] This idea that rioters form a 'mad mob', a collection of mindless 'barbarians' acting 'by instinct' and possessing the 'spontaneity, the violence, the ferocity, and also the enthusiasm and heroism of primitive beings' to quote Le Bon's own words, is challenged by Steve Reicher and Cliff Stott, two of the world's foremost experts on the psychology of crowds, whose take on the 2011 London riots we saw earlier. They believe that the 'contagion' model, used to explain why 'ordinary people' get drawn into riots, is too simplistic and ignores the fact that, as they put it, 'riots are generally comprised of several events'.[25] They advocate a more nuanced and less broad-brush search for explanations than those offered by the popular press and politicians courting popularity.

What does seem to be true, when it comes to helping explain that brief moment when impulsive action replaces reflection, is that when uncertain how best to respond we take our cue from the behaviour of those around us.

## Physical arousal and copycat behaviour

In a famous experiment from the 1960s, American psychologists Stanley Schachter and Jerome Singer gave some of their subjects a shot of adrenalin – a drug that stimulates the sympathetic nervous system – while others received a placebo.[26] Of those injected with adrenalin, one group was told what they had received while the remainder were, misleadingly, informed it was a new drug, Suproxin, designed to improve eyesight. Those told they had received adrenalin were advised it would increase their heart rate and might make them feel flushed. Those misinformed about the nature of the injection were told the drug might cause them to feel numb, itch and suffer from a mild headache. They were then asked to complete a questionnaire while waiting for the next stage of the experiment to commence. Waiting with them was an actor, employed by the researchers. He had been instructed to appear either euphoric – playfully tossing rolled up balls of paper into the rubbish bin and making paper aircraft – or angry, complaining about the stupid nature of the questions. The misinformed subjects, who found their hearts beating more rapidly due to the adrenalin but did not know *why* they felt so hyped up, tended to match the mood of the actor. When he seemed euphoric they became euphoric. When he appeared angry they too became cross. This effect was not, however, seen in either the placebo group or those who knew they had been injected with adrenalin and hence why they felt more aroused. 'Given a state of sympathetic activation, for which no immediately appropriate explanation is available, human subjects can be readily manipulated into states of euphoria, anger, and amusement,' the researchers commented. This experiment helps explain how a crowd's mood can switch from benign to furious within a few seconds. All it takes is for a few people to start sounding angry and behaving angrily, for example by yelling or throwing missiles, for others to respond in the same way. The mere presence of other people makes us feel more tense and aroused – it's part of our primitive fight or flight survival instinct – and if we are unable to explain why we feel this way then we may misattribute its cause.

In summary, the effect occurs when the arousal generated by one situation – the flimsy bridge, the scary film (Chapter 8)or being part of a large crowd – is misattributed to some other cause. Sexual attraction in the case of the bridge and the film, anger and alienation in the case of rioters. I am not, of course, suggesting that such a theory explains why people riot, only that it may well be one of the many factors leading to such impulsive and – in the case of many of those involved – uncharacteristic behaviours. Le Bon claimed that a crowd's 'most striking peculiarity' is in fact an impulse based on the notion that whenever we are in doubt about what to do next we should copy what everyone else is doing. By acting in this way we not only avoid the embarrassment of standing out from the crowd but also reduce feelings of personal responsibility for our actions. 'Generally speaking, when incidents occur, crowd members do not just watch each other watching what is happening,' report Steve Reicher and Cliff Stott. 'Crowd members will be more likely to riot when they can see from the expressions on the faces, from the tone of chants, from incipient acts of violence such as stone throwing – and then by the expressions on the faces in response to such acts – that other people are also ready and willing to take on those they see as their foes. The precipitating incidents thereby generate the sense of unity, shared purpose and of power that allows grievance to be translated into retaliation.'[27]

## 'Follow my leader' –
## the power of the mimicry heuristic

From our early years we define ourselves as a member of various social groups. At first this comprises our immediate family, where we acquire and perfect tactics for winning attention and gaining approval such as smiling, laughing, crying or even having temper tantrums. As we grow up our social circle expands to school and teachers and peer groups. Later it includes employers and colleagues at work, friends and neighbours at home. All the time we are gradually widening the social groups from whom approval is sought. Called the 'mimicry heuristic', this instinct is hard-wired into our brain and probably derives from an innate need to be accepted and viewed as 'normal' by others. It underpins our desire to join clubs, form societies, become fans or 'dedicated followers of fashion'. For those whom mainstream society deems to be outsiders, perhaps for reasons of class, colour, sexual orientation or some other minority preference, a popular option is to form their own groups, clubs and societies within which they can relax with like-minded others. The downside of the mimicry impulse is that, as we have seen in this chapter, it can lead to copycat suicides and mob-rule mentality.

## Stepping back from the brink

Just as the decision to take one's own life can arise from an impulse, so too can the decision not to do so. While working in clinical psychology, I treated a deeply depressed 23-year-old man who, following a row with his 16-year-old girlfriend, decided to jump off the cliffs at Beachy Head. Driving to his death, the young man (whom I shall call Barry) was stopped by a red signal on the last set of traffic lights before the headland. His car was in the left-hand lane and, as he waited for the lights to change, a red open-topped sports car driven by an attractive blonde girl pulled up beside him. They glanced casually at one another and she gave him a radiant smile, before the lights changed and she pulled away to make a right-hand turn. We may imagine how Barry's mirror neurons, firing in response to that bright and cheerful smile, triggered positive emotions. Perhaps these feelings then triggered memories of happier days and occasions. In any event he abruptly changed his mind about killing himself, drove home and from that point on made a rapid recovery. 'It occurred to me,' he later explained, 'that if a beautiful girl still thought I was worth a smile then I did have something to live for after all.' The impulse to self-destruct had, on an impulse, been transformed into the impulse to live. If the traffic lights had not been at red when he drove up, if the girl in her open-topped car had not arrived at that precise moment, and if she had merely stared straight ahead instead of meeting his gaze and smiling, then the outcome would almost certainly have been tragically different.

All the evidence suggests that, in many cases, the destructive impulse – whether inwardly or outwardly directed – can often be defused or diverted into more constructive behaviour. But the window of opportunity is small and the amount of patience required is often enormous. Coastguard Don Ellis, who over 20 years has managed to 'talk down' hundreds of potential suicides, knows just how thin is the dividing line between life and death on the cliff top. The longer he can engage the would-be suicide in conversation the better his chances of dissuading them from jumping. On some occasions he has sat beside a potential jumper for up to five hours, patiently listening to their problems in a calm and non-judgemental manner. Being able to distance oneself in time and/or place from any impulse makes it significantly less likely you will engage in it. This applies as much to making impulse purchases or falling in love at first sight as to participating in a riot or taking one's own life. As we shall see in the next chapter, even a brief interruption can often prove sufficient to inhibit the impulse by strengthening self-control.

# CHAPTER 12

# Deplete Us Not Into Temptation

'The capacity of the human organism to override, inter-
rupt, and otherwise alter its own responses is one of the
most dramatic and impressive functions of human self-
hood, with broad implications for a wide range of behaviour patterns.' Mark Muraven, Dianne M. Tice, and Roy
F. Baumeister, *Self-Control as Limited Resource: Regulatory
Depletion Patterns*

As Ulysses and his crew sailed home from the Trojan War they
came close to the island home of the Sirens, beautiful maidens
who lured sailors onto the rocks by their enchanted singing.
Determined to hear their wondrous songs and survive, Ulysses
ordered his men to tie him tightly to the mast and then block
their ears with wax 'so he alone could hear the Sirens' Song and
live'. The ploy worked. Legend tells how 'the Sirens' melody fell
upon Ulysses' charmed ears; but, although he commanded and
implored his men to set him free and alter their course, they kept
steadily on until no sound of magic song could reach them, when
they once more set their leader free.'[1] Ulysses was wise enough to
realise that while he might be captain of his vessel he was not
'master of his fate' when it came to irresistible song of the Sirens.
He knew he lacked the self-control to stop himself steering his
vessel onto the rocks. The paradox is that, had he succeeded in
freeing himself, his struggle to do so would have weakened rather
than strengthened his self-control.

Research has shown that the main limit on exercising

our willpower is willpower itself. By exerting self-control in one situation we diminish our capacity to do so in another. If you row with your boss or fight with your partner, you will find it considerably harder not to binge on chocolate, console yourself with a drink or light up a cigarette. When under stress, the effort to maintain control in your life makes you significantly more vulnerable to a whole host of impulses, many of them unhealthy or even life-threatening. Controlling our impulses is a lot like using our muscles. Both grow tired through use and the more they are used the more fatigued they become. In this chapter I shall be considering the key role such fatigue plays in all the different impulses described in this book, from falling in love to impulse buying, and overeating to suicide and rioting.

## What is self-control?

Self-control is an ability to modify our words and actions to achieve a desired goal. Typically this means ensuring behaviour conforms to the values, morals, ideals and expectations of social and cultural norms. While some psychologists use the terms 'self-control' and 'self-regulation' interchangeably, my own preference is to regard self-control as a conscious and deliberate System R response. It is self-control that enables us to inhibit or override one response in order to make another possible. The term 'self-regulation', by contrast, describes the automatic, homeostatic processes required to maintain the constant; for example, by ensuring our core body temperature remains very close to 37°C in the face of widely varying external conditions.

An absence of adequate self-control has been linked to a wide range of impulse-control problems, including overeating, alcohol and drug abuse, heroin addiction, crime and violence, overspending, sexually impulsive behaviour, unwanted pregnancy, smoking and some types of suicide. It is also likely to play a role in school underachievement, inability to persist at some tasks, relationship problems and dissolution, and many more. In an investigation into the number of times each day we have to exercise self-control, psychologist Wilhem Hofmann, from the University of Chicago Booth School of Business, supplied 205 adults with bleepers.[2] Over a period of a week, Hofmann and his colleagues used these to contact participants at random intervals and ask whether they had experienced any temptations. If so, how strong were these desires? Had they succumbed or resisted and, if the latter, how successful had they been? Some 8,000 temptations were recorded over the seven-day period. How easy they were to resist depended not only on the intensity of the desire but whether or not they conflicted with other, higher priority goals. A woman who wanted to wear a bikini on holiday in a few weeks' time might find saying no to a cream doughnut easier than one with no such holiday plans. Personality also played a role in the ability to exercise effective self-control, as did situational and interpersonal factors.[3] These included the influence of alcohol, the presence of other people and, especially, the presence of others who already had succumbed to temptation. As the old song says, when 'everyone is doing it, doing it, doing it' exercising self-control becomes far harder, as we saw when considering both the impulse to suicide and the impulse to loot and vandalise during a riot.

## Impulsivity and self-control

In a classic experiment, Stanford University psychologist Walter Mischel investigated willpower in 4-year-olds by means of marshmallows.[4] In his study, a child would be seated in front of a table on which was a bell and an upside-down cake tin. The experimenter explained he had to leave the room but would come back immediately if the child rang the bell. He then lifted the cake tin to reveal three small marshmallows, two placed side by side and the third a short distance away. The child was told he or she could have two marshmallows on the experimenter's return but would only get one if the bell was sounded to bring him back sooner. The experimenter then left the room and hidden observers timed how long the children were willing to wait before giving up and eating the treat. Some were so impulsive they gobbled down the single marshmallow almost immediately. Others waited patiently for up to 20 minutes before he returned and they could claim the other marshmallows. Researchers found that the length of time a child was able to wait before eating the marshmallow accurately predicted future academic and personal success. By the age of 18 those who, as 4-year-olds, had been unable to wait – termed 'grabbers' – suffered from low self-esteem and were viewed by others as stubborn, prone to envy and easily frustrated. Those who had, 14 years earlier, refrained from eating the marshmallow until Walter Mischel returned – the group he named 'waiters' – demonstrated better coping skills, greater social competence and dependability. They were more assertive, trustworthy and academically successful, scoring some 210 points higher on Scholastic Aptitude Tests at the end of secondary education.

Underlying the challenge facing those children was a principle known as *temporal discounting*. To a 4-year-old, the prospect of enjoying a marshmallow immediately could have a higher value than receiving two of them later. As adults, temporal discounting lies behind the willingness of normally sensible, rational, individuals to put their long-term health, happiness and finances at risk in order to enjoy immediate gains. That Apple iPad or plasma television is far more immediately rewarding than clearing your credit-card debts. Watching a nearby diner tucking into a calorie-rich dessert may make the prospect of ordering some yourself significantly more appealing than watching your weight drop a few weeks from now. The ability to overcome such temptations and exercise self-control in adulthood confers many benefits. Those who do so, research has shown, are more popular and successful. They enjoy healthier, more stable relationships, commit fewer crimes and suffer from less mental illness.[5][6]

So why do some people find it so difficult to exercise self-control most of the time while others find difficulty in doing so only some of the time? The limiting factor, according to some psychologists, is willpower itself! 'Self-control resembles a muscle in more ways than one,' says Roy Baumeister. 'Not only does it show fatigue, in the sense that it seems to lose power right after being used, it also gets stronger after exercise.'[7] In other words, the more we have to employ self-control in one area of life the less there is available for use in another. Such a finding raises interesting questions about what exactly it is that is being depleted. It also invites a consideration of what we mean by free will and whether, in fact, we possess any such thing. If our ability to resist an impulse to overeat, over-shop, fall in love or kill ourselves depends solely on having a sufficient quantity of some specific resource, then humans can no more exercise self-control than a car can carry on motoring once the petrol tank is empty.

## Impulses vs self-control

The idea that a conflict exists between impulses and self-control, between 'passion' and 'reason', as the Greek philosophers put it, has a long history. According to Plato's *Protagoras*, Socrates argued that weakness of will cannot exist since no one would willingly act against his or her better judgement. Rather, he suggested that someone whose self-control was poor simply lacked the knowledge and 'proper perspective' of a fully informed and wise individual. In his *Nicomachean Ethics*, by contrast, Aristotle commented that people may act against their better judgement and interests when overwhelmed by their passions. In an example that is as relevant today as it was 2,000 years ago, he explained how when offered tasty but unhealthy food we might yield to temptation if our 'passion' for the treat prevented us from hearing the 'voice of reason'.

The 'muscle' of self-control, however, does more than keep our impulses in check. It forms part of a much larger collection of executive functions concerned with self-monitoring, coping with stress, considering different options, weighing up alternatives and making decisions. All of these draw on the same limited energy source, which means that resisting temptation in one direction can make it far harder to do so in another. In a series of studies led by Kathleen Vohs of the University of Minnesota[8], participants instructed to make choices among a wide range of consumer products subsequently showed a lower tolerance for pain and were prepared to drink less of an unpleasantly flavoured drink, despite being offered payment for doing so. In a related study, college students who had to choose from various college courses did less studying for a maths test, opting instead to play video games or read magazines. Other researchers have shown that when the 'mental muscle' of self-control is depleted by overuse, people are less good at logical reasoning and intelligent thought or at coping with unexpected setbacks.

## Body states and self-control

Hunger, thirst, fatigue, addiction and sexual arousal all create the desire to satisfy a physical need, by eating, drinking, sleeping, taking drugs or having sex. Called 'visceral drives', they can lead to impulses which undermine long-term goals. The most significant self-control problems, obesity (hunger), drug addiction (drug craving) and infidelity (sexual arousal), are rooted in visceral drives.[9] Theory suggests that when someone is in a 'cold', non-visceral state, temptations will strengthen self-control and long-term interests will prevail. If, however, that person is in a 'hot', visceral state, temptation will give rise to impulses designed to satisfy that need. To test this hypothesis, Loran Nordgren and Eileen Chou, of Northwestern University, recruited heterosexual males in committed relationships. One group saw a 10-minute erotic film known to produce sexual arousal[10], the second a 10-minute film of a female fashion show. The participants then examined five photographs, each depicting a young woman deemed to be 'highly attractive'. The amount of time each man studied the photographs was recorded. A week later the same two groups of volunteers returned to the laboratory and were shown the same erotic or non-erotic film and a new set of photographs. On this occasion, however, they were told that the pictures were of international students due to take classes in the psychology department the following term. By making their participants believe they might be meeting the woman in a few weeks' time, the psychologists aimed to heighten temptation. 'We predicted that non-aroused men would devote less attention to the attractive women when temptation was high (Phase 2) than when temptation was low (Phase 1),' explains Loran Nordgren, 'however, we predicted that sexually aroused men would devote more attention to the attractive women when temptation was high (Phase 2) than when temptation was low (Phase 1).' The results were in line with the theory. Men in the 'cold', non-visceral state (favouring long-term goals) showed much less interest in the photographs than those, in a 'hot', visceral state (which favours short-term goals).

## Deplete into temptation

In a study of what Roy Baumeister – partly in homage to Freud – termed 'ego depletion', two sorts of foods were placed before hungry participants. One was a tray of chocolate biscuits, freshly baked so that their aroma filled the laboratory, together with some chocolates. The other was a bowl of red and white radishes. The participants were divided into three groups. The first group was asked to eat 'at least two or three radishes' but to ignore the tempting biscuits and chocolates. The second was allowed to sample at least two or three biscuits or a handful of the small chocolate sweets. The third group was offered no food at all. The experimenter then left the room for five minutes and surreptitiously observed the participants through a one-way mirror, noting the amount of food eaten and checking whether they ate only the foods assigned to them. He then presented his subjects with a series of unsolvable problems and timed how long they struggled with them before finally giving up on the impossible task. 'Resisting temptation seems to have produced a psychic cost,' notes Roy Baumeister, 'in the sense that afterward participants were more inclined to give up easily in the face of frustration. It was not that eating chocolate improved performance. Rather, wanting chocolate but eating radishes instead, especially under circumstances in which it would seemingly be easy and safe to snitch some chocolates, seems to have consumed some resource and therefore left people less able to persist at the puzzles.'[11]

In a later study, Mark Muraven, from the University of Albany, and his colleagues instructed a group of male students who were social drinkers not to think of a 'white bear' for several minutes. All the students were then asked to take part in a 'taste test' for different beers after first being warned not to drink too much as they would shortly have to undertake a driving test.

Despite this caution those who had been instructed not to think of the 'white bear' drank significantly more than those who had received no such instruction.[12] Variations on this type of experiment have consistently supported the finding that exercising restraint in one situation reduces one's ability to exercise it in another.

In one study, participants whose self-control had been depleted favoured sweets over more healthy granola bar snacks. In others, they chose relatively trashy films rather than more intellectual or artistic movies. These preferences occurred even when selecting a film to watch in several days' time.[13] Actual consumption was also affected. Dieters ate more food when depleted than they would otherwise have done while those not on a diet remained relatively unaffected. 'The distinction is important,' points out Roy Baumeister, 'because it suggests that ego depletion does not simply increase appetites or pleasure seeking. Rather, it undermines the defences and the virtuous intentions that would otherwise guide behavior.'

In another study it was found that ego depletion caused consumers to pay more for a set of consumer goods than did a non-depleted control group.[14] In another, students who had been paid $10 for taking part in an experiment were offered the opportunity, ostensibly as part of a marketing campaign by the campus bookstore, to make actual impulse purchases. Those whose self-control had been depleted spent over ten times as much as the non-depleted participants.

The message from all these studies is a simple one. Refrain from having a fight with your partner before leaving for work and you may find it harder to control your irritation with a colleague later in the day. Refuse a second helping of strawberry cheesecake at lunch and you may find it far harder to resist a high-calorie snack in the afternoon. The less sleep you get the harder it will be to resist temptation later the following day. Especially if the day proves particularly stressful. The weakening of self-control as a result of ego depletion can be seen in virtually every case of impulse described in this book. The self-restraint needed to stick to a diet may actually make it harder to resist a tempting treat; the effort to stay cool, calm and collected when shopping in a crowded store can lessen one's resolve not to impulse-buy something. For the depressed and despairing the strain of coping with daily challenges can fatally diminish the willpower needed, to quote Hamlet's words, 'to suffer the slings and arrows of outrageous fortune'. As a result, even suicide may come to be seen as a preferable option. The mental effort required to maintain self-control when confronted by the stresses of joblessness, poverty, alienation and a lack of material possessions – in a society which equates these with status and self-worth – is likely significantly to weaken the ability to resist the excitement and opportunities for easy personal gain.

Ego depletion can also come about merely watching others exert willpower. In a recent study by psychologists at Yale and UCLA, participants were asked to put themselves in the shoes of a fictional hungry waiter or waitress, in a gourmet restaurant, who was forbidden to eat while working. They were then shown pictures of various products like watches, cars and appliances and rated how much they would be willing to pay for them. Exercising vicarious self-control led people to be willing to spend more on the consumer goods, as compared with a control group.[15]

## Two brain systems for self-control

Self-control results from the actions of two complementary executive-control systems[16]. The task of the first is to detect discrepancies between what was intended and what actually occurred. When an error is found the second system takes over to implement the desired response and inhibit the incompatible one. These two systems are located primarily in the anterior cingulate cortex and prefrontal cortex, regions of the brain I have described in earlier chapters. These regions are also responsible for such executive functions as self-monitoring, coping with stressors, making decisions and weighing alternatives, all of which draw on the same limited energy source. After exerting willpower, these systems are less sensitive to any mismatch between the actions taken and those needed to achieve desired goals. This resource is easily exhausted and, once depleted, the likelihood of self-control failure increases. Ego depletion may occur, for example, in people who have been trying to stop thinking about food all day[17] or coping with a stigmatised social identity.[18]

So far the discussion has been framed in terms of 'ego depletion' and 'willpower' which, as Roy Baumeister concedes, has an 'unappealing, Victorian reputation'[19] and seems similar to 'muscle fatigue'. While ego depletion is a well-established phenomenon, the nature of the resource that actually becomes depleted has long remained a mystery. In recent years the answer proposed by many eminent psychologists is that this resource is glucose.

## Running on empty – glucose and self-control

Although it makes up only 2 per cent of the body's mass, the brain consumes up to 75 per cent of the glucose in the blood. Increasing blood glucose levels has been shown to enhance executive processing, improve short-term memory and speed up reaction times. Conversely, low blood glucose has been associated with such problems as lack of self-control, aggression, criminality, poor

emotional control, impulsivity, inattention, difficulties in coping
with stress and giving up smoking. Just as a torch bulb grows
fainter and fainter when the battery runs down, so too, according
to this widely held view, does our self-control become weaker as
glucose levels fall.[20]

In 2007 Matthew Gailliot of Florida State University and his
colleagues reported that exercising self-control causes a reduction
in the levels of glucose in the bloodstream. They also suggested
that low levels of blood glucose after initial acts of self-control
were strongly correlated with poor self-control on subsequent
tasks. 'Even though nearly all of the brain's activities consume
some glucose, most cognitive processes are relatively unaffected
by subtle or minor fluctuations in glucose levels within the normal
or healthy range,' they reported. 'Controlled, effortful processes
that rely on executive function, however, are unlike most other
cognitive processes in that they seem highly susceptible to normal
fluctuations in glucose . . . Self-control, therefore, may be highly
susceptible to fluctuations in glucose. Indeed, indirect evidence
suggests that self-control failure may be more likely when glucose
is low or when glucose is not transported effectively from the body
to the brain.'[21] When they gave participants Kool-Aid lemonade
sweetened with either sugar (hence glucose) or Splenda (a sugar
substitute that, the researchers claim, does not increase blood
glucose) some of the ego-depletion effects were reduced with the
first but not the second drink.

Although these arguments seem compelling and their experimental findings conclusive, this viewpoint has been robustly
critiqued by other psychologists. Robert Kurzban of the University
of Pennsylvania, for example, has pointed out that for this account
to be correct two statements would have to be true.[22] The first is
that 'performing a self-control task reduces glucose levels relative
to a control task' and the second that 'performing a self-control

task reduces glucose levels relative to glucose levels before the task'. Neither of these propositions, he asserts, was demonstrated in the study by Matthew Gailliot and his colleagues. For the moment, therefore, the jury must remain out on precisely what resource is depleted as self-control becomes fatigued. So far as impulse control is concerned, the key point to bear in mind is that exercising self-restraint in one activity makes it harder to do so in another.

## Can self-control be strengthened?

Evidence suggests the answer is yes. But before explaining how it can be done, it will be instructive to look at one of the most damaging and misguided attempts by one group to impose self-control by physical force on another group against the latter's natural inclinations. It is a shameful tale of religious fervour and medical incompetence which, in the words of the late Dr Alex Comfort, resulted in 'one of the most curious and unedifying chapters in medicosexual history'.[23]

For more than three centuries, worries about one type of impulse above all others dominated the lives of many adults and ruined the lives of their children. That impulse was masturbation. Fear of the supposed consequences of what they called 'self-abuse' led parents, doctors and ministers of the church to seek ways of eliminating the practice by means that were often extremely cruel and humiliating. Until 1710 masturbation had, in most of Western Europe, generally been accepted as a natural activity. Aristotle believed it removed many ills of the mind and lightened the body, while other medical writers held the view that semen hoarded for too long turned into a poison which produced symptoms of giddiness and poor eyesight. The anonymous author of *Hippolytus Redivius* (1644) regarded it as a remedy against the dangerous allurements of women.

Only the church remained opposed to the practice, perhaps because it was pleasurable. One theological text noted that 'the intrinsic malice of pollution consists most probably in the intense sexual enjoyment and satiation of pleasure, occurring outside the legitimate bond of matrimony, which the effusion of seed produces . . . the effusion of semen would be legitimate for medical purposes if only it could be achieved without causing pleasure.' Attitudes changed dramatically after the publication, in 1710, of *Onania, or the Heinous Sin of Self-Pollution*.[24] The anonymous author quoted a Dr Baynard as warning: 'if we turn our eyes on licentious Masturbators we shall find them with meagre Jaws, and pale Looks, with feeble Hams and legs without Calves, their generative faculties weaken'd if not destroyed in the Prime of their Years; a Jest to others and a Torment to themselves.' The book's purpose, however, was less to moralise than to market a patent medicine designed to remedy the 'effects of indulgence' at half a sovereign a box. *Onania* might have passed virtually unnoticed and unread, had it not been taken up by Dr Simon Andre Tissot.

An eminent Swiss physician, Tissot was the Pope's adviser on epidemics and the author of a respected work on public health. In 1758 he published his notorious text *Onania, or a Treatise upon the Disorders Produced by Masturbation*. In it he claimed masturbation produced such serious medical disorders, as 'consumption, deterioration of eyesight, disorders of digestion, impotence . . . and insanity.' Writers of popular guides to bringing up children were quick to fuel parental fears that lack of self-control would lead to weakness, ill health and quite possibly madness. In *What a Young Wife Ought To Know*, published in 1901, mothers were warned that the 'organs used for throwing off the waste water of the system are so closely related to other parts of the body that handling them at all will hurt [the child] and make them sick . . . [K]now where they are at all hours of the day and night; be patient and prayerful in their training; teach them truth, and keep their confidence, and you will be rewarded with strong, pure boys and girls, who can look into your eyes candidly and say, "Mamma, I am free from this habit which leads to so much misery."'

A variety of aids were offered, by the church and medical profession, to parents seeking the strengthen their son or daughter's self-control. These ranged from bizarre mechanical devices, surgical interventions and even records warning of the dangers. In 1897, clergyman Dr Sylvanus Stall published *What a Young Boy Ought To Know* and produced phonograph recordings in which he warned parents and teachers that 'to prevent the repetition of the act of masturbation, and if possibly permanently to cure the victim of this vice' boys should be put into 'strait jackets, have their hands fastened behind their backs, tied to the posts of the bed, or fastened by ropes or chains to rings in the wall'. Gadgets available to parents included chastity belts for both sexes, spiked rings to be fitted around the penis, which would cause intense pain should it become erect, and boxes that delivered a powerful electric shock or sounded an alarm that alerted adults. Other doctors offered to supply red mercury ointment that would blister the penis. In 1786, a Dr S. G. Vogel advocated infibulation (placing a silver wire through the foreskin) and many such operations were performed. By the late 19th century even more radical surgery was being applied to both sexes. Dr Eyer of St John's Hospital, Ohio, reported dealing with masturbation in a young girl by first cauterising her clitoris, then burying silver wire sutures in it, which the girl immediately ripped out, and finally removing the entire organ.

Alex Comfort notes: 'The discrediting of the insanity story did not prevent professional moralists from continuing to propagate, as authoritative fact, matter from the height of the attack; ours is possibly the first generation in which anxiety over a harmless activity has not headed the list of psychologically disturbing factors in adolescent boys.'[25] Apart from such cruel and ineffective attempts to prevent impulsive behaviour, what practical steps can be taken?

As I mentioned at the end of the previous chapter, one of the simplest is to pause and reflect. Even a short hiatus, a brief distancing in time and/or space from the object of your temptation, can kill an impulse.[26] The next time you are on the point of an impulse buy, for example, step away from the display case and count slowly to ten. Better still, leave the store and go for a short walk. One of the most dreaded phrases any salesperson can hear from a potential customer is: 'I'll think about it!' They know, and all the sales manuals emphasise the point, that if the customer is allowed to 'think about it', even for a few moments, they are much less likely to buy it.

The powerful effect of the 'pause of reflection' was demonstrated in the study of chocolate eating among London office workers which I briefly described in Chapter 9. They were provided with chocolates as a reward (so they believed) for completing a survey on eating habits. These chocolates were presented in one of two ways. Some were placed in clear glass bowls on their desks, others in solid bowls at a slight distance. In the first case all they had to do to eat a sweet – which was constantly in their line of sight – was reach out and take it. In the second case they could not directly see the sweets and needed to get out of their seats and walk about six feet to where the bowl was located. The small difference in the ease with which the sweets could be consumed resulted in the 'on desk' group consuming six times more sweets than the 'away from desk' group. In this case, out of sight really did mean out of mind.

Aside from pausing for reflection and evaluating the proposal using System R rather than System I, self-control can also be strengthened by means of exercise. In much the same way that building up one's biceps involves using progressively heavier weights, so too does increasing resistance to ego depletion require practice with smaller temptations. These should be sufficiently worthwhile that they improve life in small ways, for example by controlling one's anger towards other motorists when driving or not swearing if frustrated, but not so demanding that they speed up the depletion of this finite resource. It is a good idea, for example, only to make one New Year's resolution rather than half a dozen. Limiting expenditure to just one area of change makes it more likely you will accomplish rather than abandon your goal. Making small physical changes in behaviour can also help increase self-control. In Chapter 9 I described an experiment in which I demonstrated how the amount of popcorn eaten when watching a film was significantly reduced when the audience wore oven gloves on their popcorn-picking hand. Recent research has shown that even such minor exercises make it easier to avoid ego depletion when coping with larger issues.[27]

## Religious belief and enhanced self-control

Kevin Rounding and his colleagues in the Department of Psychology at Queens University, Ontario, tested a theory that religion promotes self-control and by doing so encourages socially acceptable behaviour. This is based on the argument that religious beliefs developed in the first instance to enhance and extend control over events.[28] A belief in God makes the world seem a more orderly and predictable place.[29] This view is supported by a study[30] showing that exposure to random or uncertain events increases people's religious beliefs. As a wartime saying among soldiers put it: 'You never find an atheist in a foxhole!'

The larger the population the greater the threat of freeloading posed by social anonymity and the greater the prevalence of moralizing, all-powerful and omniscient gods.[31] 'More specifically,' says Kevin

Rounding, 'religion can provide a solution to the self-regulation dilemmas inherent in cultural life; it helps people to control selfish impulses that might harm group interests, to subordinate short-term temptations to long-term goals, to strengthen inner restraints.'[32] From the results of their study they drew two major conclusions. The first was that self-control is central to social stability and the second was that religion is an effective cultural mechanism for regulating self-control, one that enabled our early ancestors to make evolutionarily adaptive decisions despite harsh environmental challenges. 'Thus, ancestral societies may have culturally selected religious beliefs for their ability to promote self-control, which in turn is associated with a myriad of positive behaviours, ultimately facilitating social interactions and cooperation among increasingly larger numbers of people who are not biologically related.'

## Summing it all up

Where does all this research leave us in our search to understand the nature of the impulse? Four points can be made with some confidence.

First, an impulse is specific rather than general. It motivates us to undertake some particular action, usually one associated with an urgent desire to satisfy a need. Such visceral needs as hunger, thirst or sexual desire are high on the list. When in any of these 'hot' states, the strength of the impulse makes self-control far harder and less certain.

Second, self-control is a finite resource that rapidly becomes depleted, although the jury is still out on what exactly *is* depleted. Damage to certain parts of the brain, whether through illness or injury, can also fundamentally undermine a person's ability to control their impulses.

Third, an impulse is a primitive hedonic reaction to the tempting stimulus directed towards short-term gratifications. The power of the impulse does, however, quickly fade with time.

Finally, because the impulse is a product of System I thinking, most are performed so effortlessly we usually remain unaware of the behaviour generated. That is unless something unusual or unexpected happens, either when performing the action or as a consequence of it.

I would have thought nothing about my impulsive visit to the Belfast cinema and Fred Eichler is hardly likely to have remembered his impulse to stop and chat with colleagues on his way to the restroom were it not for the terrible events that followed. Similarly Tony, the boy who scrambled away from a barn fire with seconds to spare, or Peter, whose panic on a moving escalator very likely saved his life, might have recalled with slight embarrassment their moment of inexplicable dread but were unlikely to have attached much importance to it had no threat to their lives occurred. The fact that our ability to control impulses depends on factors that remain outside our control, such as the health or otherwise of specific brain regions, raises important questions about the nature of free will. In the final part of this book, therefore, I shall explain why our belief in free will, while highly improbable from a neuro-scientific viewpoint, is absolutely essential from a social one.

# Free Will Is a Grand Illusion

> 'Men are mistaken in thinking themselves free; their
> opinion is made up of consciousness of their own actions,
> and ignorance of the causes by which they are deter-
> mined. Their idea of freedom, therefore, is simply their
> ignorance of any cause for their actions.'
> Spinoza, *The Ethics*

After a lifetime of happy marriage and respectability a 40-year-old American man suddenly and inexplicably became a paedophile. He began collecting child pornography, propositioning children and making sexual advances to his 12-year-old stepdaughter. Found guilty of child molestation, he was offered a choice between serving a prison sentence or attending a sexual addiction programme. Despite a strong desire to avoid jail the man was incapable of controlling his urges. The evening before his prison sentence was due to start he walked into the emergency room of the University of Virginia Hospital and requested a brain scan. This was performed and revealed a large tumour pressing on his right orbitofrontal cortex.[1] As we have seen, this region of the brain is involved in regulating social behaviour. Damage here can impede the acquisition of moral and social knowledge and increase the risk of poor impulse control. Surgery was performed, the tumour removed and the paedophilic impulses disappeared. When, sometime later, they returned, a further scan was performed which showed that the tumour had grown back. A second operation removed the growth and the sexual desires again vanished.

Most reasonable people would, I imagine, accept that this man's tumour absolved him from responsibility for his deviant behaviour. Yet in the absence of such a tumour, or some other clear damage, they would almost certainly have condemned him as a dangerous and chronic paedophile. This despite the fact his behaviour might well have been caused by brain changes of a less visible and currently unidentified nature.[2] The unfortunate man had no more control over his behaviour than did brain-damaged Phineas Gage or the Spaniard EVR whose stories I recounted in Chapter 3. Such cases challenge the almost universally held belief in free will.

A survey, conducted in 1998, of 36 countries found over 70 per cent of those questioned were convinced they had free will. For this reason most people resist the idea that those who commit crimes are not responsible for their actions. One study found less than one person in four (36 per cent) regarded recidivistic crime as an organic disorder.[3] Yet a vast body of research reveals a high level of mental illness among prisoners. According to some estimates, the number of prisoners worldwide suffering from a serious mental illness runs into millions. In one study a quarter of defendants, evaluated for competency, were found to be medically and legally incompetent to stand trial.[4] A review of 23,000 prisoners showed them to have a significantly higher level of psychosis or major depressive illness, and to be ten times more likely to exhibit antisocial personality disorder, than the population as a whole.[5] These mental health problems are often associated with a history of childhood maltreatment or trauma.[6]

In spite of such findings, most judges, juries, lawyers and members of the public continue to regard those who come before the courts as having exercised their free will in deciding to break the law. Even when the behaviour is disruptive rather than criminal there is often a great reluctance to attribute it to brain pathology. Children with ADHD, described in Chapter 4, are far more likely to be regarded as 'naughty', 'badly behaved' or 'attention-seeking' than as suffering from a neurological problem. 'At the core of the question of free will is a debate about the psychological causes of action,' says psychologist Roy Baumeister of Florida State University. 'That is, is the person an autonomous entity who genuinely chooses how to act from among multiple possible options? Or is the person essentially just one link in a causal chain, so that the person's actions are merely the inevitable product of lawful causes stemming from prior events, and no one ever could have acted differently than how he or she actually did?'[7]

'I am the master of my fate,' claimed the 19th-century writer William Ernest Henley in his poem *Invictus,* 'I am the captain of my soul.' Two hundred years later neuroscientists, myself included, beg to differ. Over the past decade, research in psychology and neuroscience increasingly points to the fact that our belief that we have free will is without foundation.[8] 'The subjective impression that when you make a choice you really can choose any of several options is an illusion,' says Baumeister, 'because forces outside your consciousness are in motion to determine what you will choose, even if you do not know until the last minute what that choice will be.'[9] Research strongly suggests that our illusion of free will arises from unconscious brain activity.[10][11][12] There is no 'ghost in the machine'. No 'homunculus' or 'little man' lurking somewhere in our head deciding every action. There is no 'me' in 'my'. Cartesian dualism has been laid to rest. 'Minds,' in the words of Marvin Minsky, 'are what brains do.'[13]

## Conscious mind – spectator or player?

Among the first scientists to reveal that consciousness is a spectator rather than an instigator of actions was Benjamin Libet, Professor of Physiology at the University of California, San Francisco. During the early 1980s, he and his colleagues conducted what neuroscientist Susan Blackmore has described as 'the most famous experiment ever done on consciousness'.[14] By attaching electrical sensors to the scalps of volunteers, they were able to show that brain activity associated with a decision occurred in advance of an individual becoming consciously aware of making that decision.[15] Participants were asked to report the moment they had the intention of making a spontaneous movement by noting the position of a dot moving on a computer screen. Brain recordings showed that a spike in electrical activity, a so-called readiness potential (*bereitschaftspotential*) occurred about half a millisecond before they became aware of their decision to act. Libet's groundbreaking study, later successfully replicated, suggests that the following sequence of events occurs.

Stage 1 – Brain activity associated with some activity.
Stage 2 – Conscious awareness of a desire to perform that action occurs about half a second later.
Stage 3 – Action is performed.

While it may appear to us, when we perform some action, that Stage 2 precedes Stage 1, this is not the case. Only after the brain has decided on a specific act, do we become consciously aware of that decision. It is a sequence so counter to everyday experience as to seem scarcely credible. 'Common sense tells us we know when actions are ours because we have caused them,' says Daniel Wegner of Harvard University. 'We are intrinsically informed of what we do by our conscious will. But it turns out people can be mistaken about their own authorship, sometimes because they suffer from schizophrenia, dissociative disorder, or a psychogenic movement disorder – or because they encounter situations that mislead them about the origins of action.'[16]

Not only were the general public and individuals with a vested religious or philosophical interest in defending the notion of free will outraged by Libet's findings, but so too were many neuroscientists. The vitriolic opposition his experiments generated is all the more remarkable since, as neuroscientist Susan Blackmore points out: 'Most scientists claim to be materialists. That is, they don't believe that mind is separate from body, and firmly reject Cartesian dualism. This means they should not be in the least surprised by the results. Of course the brain must start the action off, of course the conscious feeling of having made it happen must be illusory. Yet the results created uproar. I can only think that their materialism is only skin deep, and that even avowed materialists still can't quite accept the consequences of being a biological machine.'[17] Despite this torrent of criticism, other researchers have not only confirmed Libet's findings but also discovered that the onset of conscious intention is even later than he suggested.[18] [19]

## Free will as an essential social construct

What effect would a widespread acceptance that there is no such thing as free will have on society? All the evidence suggests it could prove a personal and social disaster. Previous research has demonstrated that a belief we can control events in our life, that we are able to exert agency and make independent choices, strongly affects our actions and intentions, our motives and our behaviours.[20] [21] 'When people believe that they cannot exercise control over their behavior and the events that affect their lives, they perform poorly, even though they have the skills to do better,' says Davide Rigoni from the Department of Experimental Psychology at the University of Ghent.[22]

Social psychologists have shown that people who doubt they have free will are more likely to cheat[23], behave aggressively towards others and engage in a range of anti-social behaviours. A disbelief in free will has also been found to alter the way in which the brain functions at a very basic level.[24] [25] [26] But why should this be? In Tom Stoppard's play *Jumpers*, McFee, a professor of logic and a sceptic, is shot dead. Attempting to comfort those distressed by his murder, an inept police inspector by the name of Bones remarks: 'It isn't as though the alternative was immortality!' But it is precisely our yearning for immortality, in one form or another, that makes our desire to believe in self-will so powerful and so firmly entrenched. Man is the only animal to blush, speak and know, decades in advance, of the certainty of his own demise. 'Human beings are . . . by virtue of the awareness of death and their ultimate helplessness before and vulnerability to annihilation, in constant danger of being incapacitated by overwhelming anxiety,'[27] comments Sheldon Solomon from the Department of Psychology at Skidmore College, Saratoga Springs. In an effort to manage this terror, every culture offers a narrative to explain the origins of the universe, lay down rules for acceptable conduct and promise immortality through heaven (filled either with angels or virgins, depending on your doctrinal inclinations), reincarnation or – in our secular age – grand monuments, long-lasting foundations, and painting, sculpture or writing. 'These cultural modes of death transcendence,' says Sheldon Solomon, 'allow individuals to feel like they are heroic participants in a world of meaning.'[28] Central to all these defensive psychological functions is a belief in self-will. If we do not believe we (whoever 'we' actually are) control our destiny and that any fault, as Shakespeare says, 'lies not in our stars but in ourselves' then the prospect of death truly does become 'mighty and terrible'. Self-will, along with all other world views, may indeed ultimately prove to be a shared fiction but it is a fiction on which the essential truth of our life depends.

Consider too the social consequences of a rejection of any

notions of free will. In such a world, criminals would be freed from responsibility for their actions, rapists and child molesters must go unpunished, dictators would commit genocide with even fewer legal or moral sanctions than they face today. Any behaviour, no matter how cruel or inhuman, might be shrugged off by the perpetrator with a casual 'nothing to do with me, it was my brain that did it!'

So while neuroscientists and psychologists may dismiss free will as an illusion, it is an illusion individuals and society could never live without. As a social construct, free will is all that stands between an ordered society and chaos. Our *belief* that we make up our minds, possess autonomy and can exercise control over our behaviours matters far more than scientific truth. It seems to us that we have conscious will. That we are free agents. That we cause our actions and should take responsibility for what we say and do. And that everyone else should do the same. 'Although it is sobering and ultimately accurate to call all this an illusion,' says Daniel Wegner, 'it is a mistake to conclude that the illusory is trivial.'[29] This should not, however, blind us to the fact that – as impulses help demonstrate – the mind does not run the brain. The brain runs itself. The mind is part of the running.

# Notes and References

## Introduction

**1** Definitions included: 'behaviour with no thought whatsoever' (Stanford, M. S. & Barratt, E. S. (1992). 'Impulsivity and the multi-impulsive personality disorder'. *Personality and Individual Differences* 13: 831–4); 'action or instinct without recourse to ego restraint' (English, H. (1928). *A Student's Dictionary of Psychological Terms.* Yellow Springs OH: Antioch Press); 'swift action of mind without forethought or conscious judgement' (Demont, L. (1933). *A Concise Dictionary of Psychiatry and Medical Psychology.* Philadelphia: Lippincott); 'human behaviour without adequate thought' (Hinslie, L. & Shatzky, J. (1940). *Psychiatric Dictionary.* New York: Oxford University Press); 'a sudden and strong desire, especially desires originating in the Id' (Smith, L. (1952). *A Dictionary of Psychiatry for the Layman.* London: Maxwell); 'the failure to resist an impulse, drive or temptation that is harmful to oneself or others' (Sutherland, S. (1989). *Macmillan Dictionary of Psychology.* London: The Macmillan Press); 'behaviours or responses that are poorly conceived, premature, inappropriate, and that frequently result in unwanted or deleterious outcomes' (Greenberg, J. & Hollander, E. (2003). 'Brain function and impulsive disorders'. *Psychiatric Times* 1 March 2003).

**2** Exactly who or what the devil actually was changed over the years, as did the names by which he was known. 'In addition to Satan, Beelzebub and . . . the Evil One [he] is called in the New Testament the prince of this world, the great dragon, the old serpent, the prince of the devils, the prince of the power of the air, the spirit that now worketh in the children of disbelief, the Antichrist,' says author Paul Carus. 'Satan is represented as the founder of an empire that struggles with and counteracts the kingdom of God upon earth.' See: Carus, P. (1969). *The History of the Devil.* New York: Lands End Press.

**3** Sprenger, J. & Henricus Institoris (1486/1968). *Malleus Maleficarum.* London: Folio. 18–19.

**4** In 1484, Pope Innocent VIII issued a papal bull – *Summis desiderentes affectibus* – in which he warned of a diabolic conspiracy involving 'many persons of both sexes' who had 'abandoned themselves to devils, incubi and succubae, and by their incantations, spells, conjurations, and other accursed charms and crafts, enormities and horrid offences . . . do not shrink from committing and perpetuating the foulest abominations and filthiest excesses to the deadly peril of their own souls . . . and are a cause of scandal and danger to very many'.

**5** McCown, W. G. & DeSimone, P. A. (1993). 'Impulses, impulsivity, and impulsive behaviors: a historical review of a contemporary issue'. In *The Impulsive Client: Theory, Research, and Treatment.* Washington D.C.: American Psychological Association. 8.

**6** Theopholi Bonet, was born in Geneva on 15 March 1620, the son and grandson of physicians. For a time he was physician to the Duke of Longueville but deafness forced him to give up his practice and to confine himself to writing. He died on 29 March 1689. In 1700 his book *Sepulchretum* [The Cemetery] was published in Paris by Cramer & Perachon. It's full title was: *Theopholi Boneti Medicinae Doctoris,*

*Sepulchretum Sive Anatomia Practica, ex Cadaveribus Morbo Denatis, Proponens Historias et Observationes Omnium Humani Corporis Affectuum, ipsorumq; causus reconditas revelans. Quo Nomine tam Pathologiae Genuinae, quam Nosocomiae Orthidoxae Fundatrix, imo Medicinae Veteris ac Novae Promptuarium dici meretur. Cum Indicibus necessariis. Editio Altera, Quam Novis Commentariis.* Paris.

7 Philippe Pinel (1745–1826) was a French physician who was instrumental in the development of a more humane approach to caring for the mentally ill. In his 1801 book *Traité médico-philosophique sur l'aliénation mentale; ou la manie,* Pinel discusses a psychologically oriented approach to treatment. Translated into English in 1806, under the title *A Treatise on Insanity,* it had an enormous influence on both French and Anglo-American psychiatrists during the 19th century.

8 Born on a farm near the town of Normal, Illinois on 1 May 1869, Walter Dill Scott applied psychology to various branches of commerce, including personnel selection and advertising. In 1903 he wrote *The Theory and Practice of Advertising,* the first book on the topic, and in 1909 became Professor of Applied Psychology and Director of the Bureau of Salesmanship Research at Pittsburgh Carnegie Technical University. He died in 1955.

9 Hirt, E. (1905) *Die Temperamente, Ihr Wesen, Ihre Bedeutung Fur Das Seelische Erleben Und Ihre Besonderen Gestaltungen.* Leipzig: Barth.


## Chapter 1: The Impulse That Saved My Life

1 On Monday 9 August 1971, British Forces together with the largely Protestant RUC (Royal Ulster Constabulary) launched Operation Demetrius. This involved the arrest and detention

without trial of anyone accused of belonging to a paramilitary organisation. It was the start of internment. Over the next four years, 342 men and youths – almost all of them Catholics – were detained and 11 killed. Internment, which lasted until 1975, was introduced on the orders of Northern Ireland's Prime Minister, Brian Faulkner. Special Branch and MI5 drew up a list of 450 suspected terrorists but only 350 of these were ever tracked down and interned. The British urged Faulkner to include some suspected Loyalist paramilitaries on the list but, it is claimed, he flatly refused to do so and not one of them was ever interned.

2 'The Men Behind the Wire' was released on 14 December 1971 by Release Records in Dublin. With its opening refrain 'Armoured cars and tanks and guns, came to take away our sons. But every man will stand behind the men behind the wire', it went straight into the Irish charts. It remained there for months, outselling every other single released in Ireland up to that point. In late January 1972 it was number one in the Irish charts, where it remained for three weeks, with royalties being donated to families of the internees. McGuigan was arrested and held without charge in a later round of internment. Many in Ireland regarded this as an act of revenge by the British government.

3 Some of my Belfast photographs can be found on www.dlpl.org.

4 Kelly's Cellars was the meeting place for Republican hero Henry Joy McCracken and the United Irishmen when they were planning their revolt against British rule in 1798.

5 A report by the Police Ombudsman in February 2011 found an 'investigative bias' by police during the original investigation into the atrocity. The Ombudsman accused Royal Ulster Constabulary officers of being so blinkered by their belief the explosion was caused by an IRA bomb in transit they failed to investigate the

bombing properly and no one was ever charged. Today a simple memorial listing the names of all those who died marks the spot where the pub once stood.

**6** This account is based on a number of reports by survivors, especially an excellent and moving article, '9/11 Survivors of the Twin Towers', by Glenda Cooper, published in the *Daily Mail* on 14 October 2011.

**7** The reason why Tony consulted me at the age of 45 was that following the death of his father he had developed panic attacks at work. When his father was dying of cancer, the hospital had promised to let him and his brother know should the old man's condition worsen. In the event, however, they failed to make the call and when he arrived that evening his father was already dead. Tony had always felt a deep sense of guilt that he had never been able to apologise for the damage he and his brother had caused to the farm, damage which almost led to his father being dismissed from his job. Following the disaster neither parent had spoken to the boys about the accident and, indeed, they refused to ever discuss it. He had wanted to make his apologies before his father died and felt an overwhelming sense of guilt when the hospital's oversight prevented him from doing so.

**8** *Belfast Telegraph* Friday 19 August 2011.

**9** James, W. (1890/1950). *The Principles of Psychology* 2. New York: Dover Publications Inc. 542.

## Chapter 2: Impulses and Your Zombie Brain

**1** Atkinson, A. P., Thomas, M. & Cleeremans, A. (2000). 'Consciousness: Mapping the theoretical landscape'. *Trends in Cognitive Sciences* 4: 375.

**2** Solomons, L. & Stein, G. (1896). 'Normal motor automation'. *Psychological Review* 36: 492–572. There is a famous anecdote about the relationship between William James and Gertrude Stein, his young student. In her 1995 book *Favoured Strangers: Gertrude Stein and Her Family* (New Brunswick, NJ: Rutgers University Press), Linda Wagner-Martin recounts how, not feeling like taking a philosophy exam James had set, Stein wrote at the top of her paper: 'Dear Professor James, I am so sorry but really I do not feel a bit like an examination paper in philosophy today.' She then walked out. The next day she had a postcard from William James saying: 'Dear Miss Stein, I understand perfectly how you feel: I often feel like that myself.' Underneath this message he gave her work the highest mark in his course. Her angry fellow students protested so loudly she was eventually obliged to sit the exam. Stein was only awarded a C grade on this occasion.

**3** Lakoff, G. & Nunez, R. E. (2000). *Where Mathematics Comes From: How the Embodied Mind Brings Mathematics into Being.* New York: Basic Books. 27.

**4** Michael Posner and C. R. R. Snyder called the slower, reflective method 'conscious processing' and the fast, non-conscious one 'automatic activation'. (Posner, M. & Snyder, C. R. R. (1975). 'Attention and cognitive control'. In Soloso, R. L. (ed.) *Information Processing and Cognition: The Loyola Symposium.* New York: Wiley. 55–8.) Other psychologists have named them 'associative vs. rule based'; 'input modules vs. higher cognition'; 'explicit vs. implicit'; 'impulsive vs. reflective'. The terms System 1 and System 2 have also been widely used to provide more neutral forms of description. For the purposes of this book I shall follow the nomenclature suggested by Professors Fritz Strack and Roland Deutsch of the Department of Psychology at the University of Würzburg, by referring to them as impulsive and reflective. (Strack, F. & Deutsch, R. (2004).

'Reflective and impulsive determinants of social behavior'. *Personality and Social Psychological Review* 8 (3): 221.) There is a widely held view that these two systems are separate, have distinct evolutionary histories, and are implemented by different brain regions. This idea is not, however, without its critics. Some neuroscientists and cognitive psychologists argue that rather than making a categorical distinction between the two systems we should view them as opposing ends of a continuum of different processing styles. While this is a view to which I too subscribe, describing them as separate does not result in any loss of understanding and greatly facilitates the presentation of these ideas.

5 Sellers, P. (2002). 'Something to prove: Bob Nardelli was stunned when Jack Welch told him he'd never run GE. "I want an autopsy!" he demanded'. *Fortune* 24 June.

6 The first Lorenz machine, the SZ40 (SZ for *Schlüsselzusatz*, meaning 'cipher attachment'), was introduced on an experimental basis in 1940. Enhanced versions swiftly followed. In the middle of 1942 the SZ42 came into general service to enable top-secret communications between the German High Command in Berlin and Army commands throughout occupied Europe. The Nazis remained completely confident that their cipher was unbroken and their secrets secure. They were wrong. An excellent description can also be found in the Special Fish Report by Albert W. Small at: http://www.codesandciphers.org.uk/documents/small/smallix. HTM.

7 Hinsley, F. H. & Stripp, A. (1993). *Codebreakers: The Inside Story of Bletchley Park.* Oxford: Oxford University Press. 141–66. An article on the making and breaking of the Lorenz cipher, by Tony Sale, can be found at: http://www.codesandciphers.org.uk/lorenz/fish.htm

**8** Langer, E. (1989). *Mindfulness*. Cambridge, MA: Da Capo Press: 43.

**9** Ibid.:15.

**10** Langer, E., Blank, A. & Chanowitz, B. (1978). 'The mindlessness of ostensibly thoughtful action: The role of placebic information in interpersonal interaction'. *Journal of Personality and Social Psychology* 36: 635–42.

**11** Langer et al. (1989) *Mindfulness* op. cit. 14–15.

**12** Ibid.: 22.

**13** Swiss crystallographer Louis Albert Necker first spotted this illusion in 1832 while studying drawings of various crystals.

**14** Adamson, R. E. (1952). 'Functional fixedness as related to problem solving: a repetition of three experiments'. *Journal of Experimental Psychology* 44: 288–91.

**15** Dickman, S. J. (1990). 'Functional and dysfunctional impulsivity: personality and cognitive correlates'. *Journal of Personality and Social Psychology* 58 (1): 95–102.

**16** Ibid.

**17** Herbert, W. (2010). 'On second thought', *Denver Post*, 28 October.

## Chapter 3: Inside the Impulsive Brain

**1** See, for example: *That Elusive Spark*, a play by Janet Munsil; 'The Ballad of Phineas P. Gage' by Crystal Skillman, music by Joshua Goodman (http://www.youtube.com/watch?v=f7066QRhWmM); the song 'Phineas Gage' by Dan Under (http://www.youtube.com/watch?v=3vndKirATAg). See also: Macmillan, M. (2002). *An Odd*

*Kind of Fame. Stories of Phineas Gage.* Cambridge, MA: MIT Press.

**2** The word is most likely derived from the French *tampon* or *tapon* meaning a plug or a bung.

**3** Harlow, J. M. (1868). 'Recovery from the passage of an iron bar through the head'. *Publications of the Massachusetts Medical Society* 2: 327–47.

**4** I have based my account of Phineas Gage on a range of primary and secondary sources. One of the most valuable of these is the 2002 book by Malcolm Macmillan mentioned in note (1) above. A Professorial Fellow in the Department of Psychology at the University of Melbourne, Macmillan has produced a superbly researched work that should be compulsory reading for anyone interested in the case. Not only has he accessed a wide range of original documents, but he also presents a robust critique of the ways in which Gage's story has been selectively quoted over the years by those with a particular opinion on brain localisation to promote. Gage's story, he writes, 'illustrated how easily a small stock of facts can be transformed into popular and scientific myth . . . used to support particular theoretical positions'. Other sources are:

i) (1851). 'A most remarkable case'. *American Phrenological Journal and Repository of Science, Literature, and General Intelligence* 13: 89, column 3.

ii) Boston Society for Medical Improvement (1849). Records of Meetings (Vol.VI) Countway Library Mss, B MS b.92.2.

iii) Haas, L. F. (2001). 'Phineas Gage and the science of brain localisation'. *Journal of Neurology, Neurosurgery and Psychiatry* 71: 761.

iv) Harlow (1868) op. cit. Reprinted (1993) in: *History of Psychiatry* 4: 271–81.

v) Harlow, J. M. (1848). 'Passage of an iron rod through the head'. *Boston Medical and Surgical Journal* 39: 389–93.

vi) Harlow, J. M. (1849). Letter in 'Medical miscellany'. *Boston Medical and Surgical Journal* 39: 506–507.

vii) Jackson, J. B. S. (1849). Medical Cases (Vol.4, Cases Number 1358–1929, pp 720 and 610). Countway Library Mss, H = MS b 72.4.

viii) Jackson, J. B. S. (1870). *A Descriptive Catalogue of the Warren Anatomical Museum.* Boston, MA: Williams.

ix) Macmillan, M. (2008). 'Phineas Gage – unravelling the myth'. *The Psychologist* 21 (9) September: 828–31.

x) O'Driscoll, K. & Leach, J. P. (1998). '"No longer Gage": an iron bar through the head'. *British Medical Journal* 317: 1673–4.

5 Today both are displayed in the Warren Anatomical Medical Museum at Harvard University.

6 Broca, P. (1861). 'Perte de la parole, ramollissement chronique et destruction partielle du lobe antérieur gauche'. *Bulletin de la Société d'Anthropologie* 2: 235–8.

7 Broca, P. (1861). 'Nouvelle observation d'aphémie produite par une lésion de la moitié postérieure des deuxième et troisième circonvolutions frontales gauches'. *Bulletin de la Société Anatomique* 36: 398–407.

8 Many doctors also argued that the idea of brain localisation was no more than the discredited theory of phrenology in a new guise. Even at the end of the 19th century, phrenology was a topic of great controversy. Invented by the Viennese physician Franz Joseph Gall, it proposed that, just as the body is composed of distinct organs, each performing specific physiological functions, so too does the brain comprise different 'mental organs', each

performing a specific intellectual or emotional task. While such thinking is very much in line with conventional neuroscientific thinking about a modular brain, Gall then took his speculations one step too far. By doing so he transformed what was radical into pseudoscience. He asserted that the extent to which these brain areas were developed could be determined from the outside of the skull. The more developed a mental 'faculty' – such as wisdom or humility – the larger the 'bump' produced in the bone directly above it. While this is now rightly dismissed as nonsense it was by no means the work of a quack. Gall was the foremost neuroanatomist of his day, an outstanding clinician and the man who laid the foundations for the subsequent development of modern cognitive neuroscience. For most medical men of that period, however, even the faintest whiff of phrenology was sufficient to damn a theory in their eyes as fraud. See: Uttal, W. R. (2001). *The New Phrenology: The Limits of Localizing Cognitive Processes in the Brain.* Cambridge, MA: MIT Press.

**9** Ferrier, D. (1878). 'The Goulstonian lectures on the localisation of cerebral disease'. *British Medical Journal*: 397–442.

**10** Mataró, M., Jurado, A., García-Sánchez, C., Barraquer, L., Costa-Jussà, F. R. & Junqué, C. (2001). 'Long-term effects of bilateral frontal brain lesion 60 years after injury with an iron bar'. *Archives of Neurology* 58: 1139–42.

**11** Ibid. 1140.

**12** Ibid. 1142.

**13** Cardinal, R. N., Parkinson, J. A. & Everitt, B. J. (2002). 'Emotion and motivation: the role of the amygdala, ventral striatum, and prefrontal cortex'. *Neuroscience and Biobehavioural Reviews* 26: 321–52.

**14** Whalen, P. J., Kagan, J., Cook, R. G., Davis, C. F., Hackjin, K., Polis, S., McLaren, D. G., Somerville, L. H., McLean, A. A.,

Maxwell, J. S. & Johnstone, T. (2004). 'Human amygdala responsivity to masked fearful eye whites'. *Science* 306: 2061.

**15** Benson, D. F. & Blumer, D. (eds.) (1975). *Psychiatric Aspects of Neurologic Disease.* New York: Grune & Stratton Inc. 158.

**16** Wallace, J. F. & Newman, J. P. (1990) 'Differential effects of reward and punishment cues on response speed in anxious and impulsive individuals'. *Personality and Individual Differences* 11 (10): 999–1009.

**17** Corr, P. J., Pickering, A. D. & Gray, J. A. (1995). 'Personality and reinforcement in associative and instrumental learning'. *Personality and Individual Differences* 19: 47–71.

**18** Pickering, A. D. & Gray, J. A. (1999). 'The neuroscience of personality'. In Pervin, L. A. & John, O. P. (eds.) *Handbook of Personality: Theory and Research.* 2nd edn. New York: Guilford Press. 277–99.

**19** Gray J. A. (1973). 'Causal theories of personality and how to test them'. In Royce, J. R. (ed.) *Multivariate Analysis and Psychological Theory.* London: Academic Press. 409–63.

**20** Eslinger, P. J. & Damasio, A. R. (1985). 'Severe disturbance of higher cognition after bilateral frontal lobe ablation: patient EVR'. *Neurology* 35: 1731–41.

**21** Biran, I. & Chatterjee, A. (2004). 'Alien hand syndrome'. *Archives of Neurology* 61: 292–4.

**22** Lhermitte, F. (1983). '"Utilisation behaviour" and its relation to lesions of the frontal lobes'. *Brain* 106 (2): 237–55.

**23** After expelling Tom Brown (1663–1704), Dr John Fell, Dean of Christ Church (1625–86) offered to revoke his expulsion if he could translate the 33rd Epigram of Martial: *Non amo te, Zabidi, nec possum dicere quare: Hoc tantum possum dicere non amo te.* The mocking verse was the young student's response.

**24** Gigerenzer, G. & Todd, P. M. (1999). *Simple Heuristics That Make Us Smart.* Oxford: Oxford University Press. 21.

**25** Wright, R. (2004). *A Short History of Progress.* Edinburgh: Canongate Books Ltd. 55. Archaeologists date the first fully established civilisations to around 3,000 BCE, in Sumer and Egypt. The rise of civilisations is considered by many to have started around 12,000 years ago with the first cultivation of plants.

**26)** Shermer, M. (2012). 'What we don't know'. *Nature* 484 April: 447.

# Chapter 4: The Teenage Brain – A Work in Progress

**1** Spear, L. P. (2000). 'The adolescent brain and age-related behavioural manifestations'. *Neuroscience and Biobehavioural Reviews* 24: 417–63.

**2** Patton, G. C., Coffey, C., Sawyer, S. M., Viner, R. M., Haller, D. M., Bose, K., Vos, T., Ferguson, J. & Mathers, C. D. (2009). 'Global patterns of mortality in young people: a systematic analysis of population health data'. *Lancet* 374: 881–92.

**3** Tarter, R. E., Ridenour, T. A., Reynolds, M., Mezzich, A., Kirisci, L. & Vanyukov, M. (2003). 'Neurobehavioral disinhibition in childhood predicts early age at onset of substance abuse disorder'. *American Journal of Psychiatry* 160: 1078–85.

**4** Lewis, D. (1978). *The Secret Language of Your Child.* London: Souvenir Press.

**5** Silverman, I. W. & Ragusa, D. M. (1990). 'Child and maternal correlates of impulse control in 24-month-old children'. *Genetic, Social and General Psychology Monographs* 116: 435–73.

**6** Rothbart, M. K. (1988). 'Temperament and the development of inhibited approach'. *Child Development* 59: 1241–50.

**7** Carey, W. B., Fox, M. & McDevitt, S. C. (1977). 'Temperament as a factor in early school adjustment'. *Pediatrics* 60: 612–24.

**8** Daruna, J. H. & Barnes, P. A. (1993). 'A neurodevelopmental view of impulsivity'. In McCown, W. G., Johnson, J. L. & Shure, M. B. (eds.) *The Impulsive Client: Theory, Research and Treatment.* Washington D.C.: American Psychological Association. 25.

**9** Zuckerman, M. (1991). *Psychobiology of personality*, Cambridge: Cambridge University Press. 3–4.

**10** Daruna & Barnes (1993) op.cit. 26.

**11** Silverman & Ragusa (1990) op. cit.

**12** Olson, S. L., Bates, J. E. & Bayles, K. (1990). 'Early antecedents of childhood impulsivity: The role of parent-child interaction, cognitive competence, and temperament'. *Journal of Abnormal Child Psychology* 18: 176–83.

**13** Miller, G. E., Lachman, M. E., Chen, E., Gruenewald, T. L., Karlamangla, A. S. & Seeman, T. E. (2011). 'Pathways to resilience: Maternal nurturance as a buffer against the effects of childhood poverty on metabolic syndrome at midlife'. *Psychological Science* 22 (12): 1591–9.

**14** Polanczyk, G., de Lima, M. S., Horta, B. L., Biederman, J. & Rohde L. A. (2007). 'The worldwide prevalence of ADHD: a systematic review and metaregression analysis'. *American Journal of Psychiatry* 164: 942–8.

**15** Faraone, S. V., Biederman, J. & Mick, E. (2006). 'The age-dependent decline of attention deficit hyperactivity disorder: a meta-analysis of follow-up studies'. *Psychological Medicine* 36: 159–65.

**16** Swanson, J. M., Sergeant, J. A., Taylor, E., Sonuga-Barke, E. J., Jensen, P. S., & Cantwell, D. P. (1998). 'Attention-deficit hyperactivity disorder and hyperkinetic disorder'. *Lancet* 351: 429–33.

**17** Side effects include: for methylphenidate, insomnia, nervousness and headache; for atomoxetine, decreased appetite, headache, sleepiness, abdominal pain, vomiting and nausea. Dexamfetamine has been linked to heart and eye problems and gastrointestinal, musculoskeletal, nervous system, renal, reproductive system, skin and vascular disorders.

**18** It is important to note that the list of side effects can change as more are added and much of what is written now will soon be out of date.

**19** Charach, A., Figueroa, M., Chen, S., Ickowicz, A., & Schachar, R. (2006) 'Stimulant treatment over 5 years: effects on growth'. *Journal of American Academic Child Adolescent Psychiatry* 45: 415–21.

**20** Lansbergen, M. M., van Dongen-Boomsma, M., Buitelaar, J. K. & Slaats-Willemse, D. (2010). 'ADHD and EEG-neurofeedback: a double-blind randomised placebo-controlled feasibility study'. *Journal of Neural Transmission* 31 October.

**21** Interview with Dr Jay Giedd on NIMH website, entitled 'Development of the Young Brain', 2 May 2011. http://www.nimh.nih.gov/media/video/giedd.shtml

**22** Stephen Wood quoted in Smith, D. (2009) 'The wonder of the teen brain'. *The Age* 30 March.

**23** Giedd, J. N. (2007). 'The teen brain: insights from neuroimaging'. *Journal of Adolescent Health* 42 (4): 335–43.

**24** Smith (2009) op. cit.

**25** McAnarney, E. R. (2008). 'Adolescent brain development: Forging new links?'. Editorial in *Journal of Adolescent Health* 42 (4) April.

**26** Steinberg, L. & Scott, E. S. (2003). 'Less guilty by reason of adolescence: Developmental immaturity, diminished responsibility, and the juvenile death penalty'. *American Psychologist* 58 (12): 1009–18.

**27** Steinberg, L. (2004). 'Risk taking in adolescence: What changes, and why?'. *Annals of the New York Academy of Sciences* 1021: 51–8.

**28** 'When enough is too much: Perspectives on emotions and behaviors that go too far'. Theme programme at the APS 22nd Annual Convention in Boston, 27–30 May, 2010.

**29** Steinberg, L. (2007). 'Risk taking in adolescence. New perspectives from brain and behavioural studies'. *Current Directions in Psychological Science* 16 (2). 55–9.

**30** Quoted in Deborah Smith, 'The wonder of the teen brain', The Age Education Resource Centre. 30 March 2009 http://education.theage.com.au/

## Chapter 5: Impulse and the Senses

**1** Lewis, D. (1985). *Loving and Loathing*. London: Constable & Co. 48.

**2** Lewis, D. (1985) op.cit. 50.

**3** Stoddart, D. M. (1991). *The Scented Ape: The Biology and Culture of Human Odour*. Cambridge: Cambridge University Press. 49–62.

**4** MacFarlane, A. (1975). 'Olfaction in the development of social preferences in the human neonate'. *Ciba Foundation Symposium* 33: 103–17.

**5** The record for the most acute sense of smell in the animal kingdom is held by the male emperor moth (*Eudia pavonia*), which can detect pheromones from a virgin female at a distance of seven miles, a feat made all the more remarkable by the fact that it is manufactured from an alcohol of which only 0.0001 mg is present in the female's body.

6 Fuller, G. N. & Burger, P. C. (1990). 'Nervus Terminalis (cranial nerve zero) in the adult human'. *Clinical Neuropathology* 9 (6): 279–83.

7 Von Bartheld, C. S. (2004). 'The terminal nerve and its relation with extrabulbar "olfactory" projections: lessons from lampreys and lungfishes'. *Microscopic Research and Technique* 65 (1–2): 13–24.

8 One of the chief reasons for their persecution was the fact they are pacifists, forbidden by their religious belief from participating in any type of military activity, wearing uniforms (such as those of a policeman of soldier) or paying taxes that contribute to military spending. This withdrawal from many aspects of civil authority extends to a self-imposed decree that discourages all forms of contact with non-believers.

9 Morgan, K., Holmes, T. M., Schlaut, J., Marchuk, L., Kovithavongs, T., Pazderka, F. & Dossetor, J. B. (1980). 'Genetic variability of HLA in the Dariusleut Hutterites. A comparative genetic analysis of the Hutterites, the Amish, and other selected Caucasian populations'. *American Journal of Human Genetics* 32 (2): 246–57.

10 Ober, C., Weitkamp, L. R., Cox, N., Dytch, H., Kostyu, D. & Elias, S. (1997). 'HLA and mate choice in humans'. *American Journal of Human Genetics* 61 (3): 497–504.

11 Herz, R. (2007). *The Scent of Desire*. New York: HarperCollins. 126–8.

12 Sukel, K. (2012). *Dirty Minds*. New York: Free Press. 88.

13 Koyama, M., Saji, F., Takahashi, S., Takemura, M., Samejima, Y., Kameda, T., Kimura, T. & Tanizawa, O. (1991). 'Probabilistic

assessment of the HLA sharing of recurrent spontaneous abortion couples in the Japanese population'. *Tissue Antigens* 37 (5): 211–17.

**14** Laitinen, T. (1993). 'A set of MHC haplotypes found among Finnish couples suffering from recurrent spontaneous abortions'. *American Journal of Reproductive Immunology* 29 (3): 148–54.

**15** Wedekind, C., Seebeck, T., Bettens, F. & Paepke, A. J. (1995). 'MHC-dependent mate preferences in humans'. *Proceedings of the Royal Society of London* 260 (1359): 245–9.

**16** Li, W., Moallem, I., Paller, K. A. & Gottfried, J. A. (2007). 'Subliminal smells can guide social preferences'. *Psychological Science* 18 (12): 1044–9.

**17** Tybur, J. M., Bryan, A. D., Magnan, R. E. & Hooper, A. E. C. (2011). 'Smells like safe sex: Olfactory pathogen primes increase intentions to use condoms'. *Psychological Science* 22 (4): 478–80.

**18** Not taking kindly to being scorned, the wives retaliated by hunting down and slaughtering every man they could find!

**19** Benton, D. (1982). 'The influence of androstenol – a putative human pheromone – on mood throughout the menstrual cycle'. *Biological Psychology* 15 (3–4): 249–56.

**20** Baron, R. A. (1981). 'Olfaction and human social behaviour: Effects of a pleasant scent on attraction and social perception'. *Personality and Social Psychology Bulletin* 7 (4): 611–16.

**21** Kotler, P. (1973–4). 'Atmospherics as a marketing tool'. *Journal of Retailing* 49 (4): 48–64.

**22** Lewis, D. (2012). *Retail Atmospherics: A Practical Guide to Serving Your Customers Right in the 21st Century*. Free download from www.themindlab.org.

**23** Morrison, M. (2002). 'The power of music and its influence on international retail brands and shopper behaviour: a multi-case study approach'. Australia and New Zealand Marketing Academy Conference 2001.

**24** Yalch, R. & Spangenberg, E. (1990). 'Effects of store music on shopping behavior'. *Journal of Services Marketing* 4: 31–9.

**25** Yalch, R. & Spangenberg, E. (1993). 'Using store music for retail zoning: a field study'. *Advances in Consumer Research* 20 (eds. Leigh McAlister & Michael L. Rothschild) Provo, UT. Association for Consumer Research: 632–6.

**26** Rolls, E. T., Grabenhorst, F. & Parris, B. A. (2008). 'Warm pleasant feelings in the brain'. *Neuroimage* 41 (4): 1504–13.

**27** Fiske, S. T., Cuddy, A. J. C. & Glick, P. (2007). 'Universal dimensions of social cognition: warmth and competence'. *Trends in Cognitive Sciences.* 11 (2): 77–83.

**28** Jzerman, H. I. & Semin, G. R. (2009). 'The thermometer of social relations'. *Psychological Science* 20 (10): 1214–20.

## Chapter 6: The Power of the Visual

**1** Pareidolia is not restricted to the visual sense. I once interviewed a group of educated and apparently sane men and women who claimed to hear voices from beyond the grave via a primitive radio receiver. This phenomenon, known as 'Raudive voices' after the Latvian parapsychologist Konstantin Raudive who discovered it, is said to occur in electronically generated sounds. To me it merely sounded like bursts of static!

**2** The son of a merchant, Victor Kandisky was born in Siberia on

March 24, 1849. He completed his medical training in 1872 and practiced as a general physician in a Moscow hospital for four years until he became psychotic and was confined to a mental hospital. After being discharged he worked as a psychiatrist and wrote his first book, based on his experiences. Kandinsky described several psychopathological symptoms included in the term 'mental automatism', including telepathy, reading and broadcasting thoughts, enforced speaking, and enforced motor movements. The latter syndrome was later called the Kandinsky-Clerambault syndrome. His chief contributions to psychiatry were in the fields of psychopathology, psychiatric classification, and forensic psychiatry. In September 1889, feeling the return of his psychosis he killed himself.

3 Whitson J. A. & Galinsky, A. D. (2008). 'Lacking control increases illusory pattern perception'. *Science* 322 (5898): 115–17.

4 Lewis, D. (2006). *The Man Who Invented Hitler*. London: Hodder Headline. 6.

5 Our study was a replication of one conducted in 1986. See: Rozin, P., Millman, L. & Nemerof, C. (1986). 'Operation of the laws of sympathetic magic in disgust and other domains'. *Journal of Personality and Social Psychology* 50: 703–12.

6 Kelly, D. (2011). *Yuck!* Cambridge, MA: MIT Press. 65.

7 Gregory, R. (2004). 'The blind leading the sighted'. *Nature* 430 (7002): 1484.

8 Ostrovsky, Y., Meyers, E., Ganesh, S., Mathur, U. & Sinha, P. (2009). 'Visual parsing after recovery from blindness'. *Psychological Science* 20 (12): 1484–91.

9 Russell, R. (2009). 'A sex difference in facial contrast and its exaggeration by cosmetics'. *Perception* 38 (8): 1211–19.

**10** Russell, R. (2003). 'Sex, beauty, and the relative luminance of facial features'. *Perception* 32 (9): 1093–1107.

**11** Gervais, W. M. & Norenzayan, A. (2012). 'Analytic thinking promotes religious disbelief'. *Science* 336: 493–6.

**12** Zhong, C.-B. & DeVoe, S. E. (2010). 'You are how you eat: Fast food and impatience'. *Psychological Science* 21 (5): 619–22.

**13** Segall, M. H., Campbell, D. T. & Herskovits, M. J. (1966). *The Influence of Culture on Visual Perception.* Indianapolis: Bobbs-Merrill Co. Inc. 213.

**14** Masuda, T. & Nisbett, R. E. (2001). 'Attending holistically versus analytically: Comparing the context sensitivity of Japanese and Americans'. *Journal of Personality and Social Psychology* 81 (5): 922–34.

**15** Boland, J. E., Chua, H. F. & Nisbett, R. E. (2005). 'How we see it: Culturally different eye movement patterns over visual scenes'. In Rayner, K., Shen, D., Bai, X. & Yan, G. (eds.) *Cognitive and Cultural Influences on Eye Movements.* Tianjin, China: People's Press/Psychology Press. 363–78.

**16** Ibid.

**17** Camus, A. (1955). 'An Absurd Reasoning'. In O'Brien, J. (ed. & trans.) *The Myth of Sisyphus and Other Essays.* New York: Vintage Books. 94.

**18** Proulx, T. & Heine, S. J. (2009). 'Connections from Kafka: Exposure to meaning threats improves implicit learning of an artificial grammar'. *Psychological Science* 20 (9): 1125–31.

**19** Dijksterhuis, A. & Nordgren, L. F. (2006). 'A theory of unconscious thought'. *Perspectives on Psychological Science* 1 (2): 95–109.

## Chapter 7: Impulses and the Risk-Taking Personality

1 Coates, J. M. & Herbert, J. (2008). 'Endogenous steroids and financial risk taking on a London trading floor'. *Proceedings of the National Academy of Science USA* 104: 6167–72.

2 Coates, J. M., Gurnell, M. & Rustichini, A. (2009). 'Second-to-fourth digit ratio predicts success among high frequency financial traders'. *Proceedings of the National Academy of Science USA* 106: 623–8.

3 Putz, D. A., Gaulin, S. J. C., Sporter, R. J. & McBurney, D. H. (2004). 'Sex hormones and finger length. What does 2D:4D indicate?'. *Evolution and Human Behaviour* 25: 192–9.

4 Manning, J. T., Scott, D., Wilson, J. & Lewis-Jones, D. I. (1998). 'The ratio of 2nd to 4th digit length: A predictor of sperm numbers and concentration of testosterone, luteinising hormone and oestrogen'. *Human Reproduction* 1311: 3000–04.

5 Coates et al. (2009) op. cit.

6 Kondo, T., Zakany, J., Innis, W. J. & Duboule, D. (1997). 'Of fingers, toes, and penises'. *Nature* 390: 29.

7 Bailey, A. A. & Hurd, P. L. (2005). 'Finger length ratio (2D:4D) correlates with physical aggression in men but not in women'. *Biological Psychology* 68 (3): 215–22.

8 Roiser, J. P., de Martino, B., Tan, G. C. Y., Kumaran, D., Seymour, B., Wood, N. W. & Dolan, R. J. (2009). 'A genetically mediated bias in decision making driven by failure of amygdala control'. *Journal of Neuroscience* 29: 5985–91.

9 Csatho, A., Osvath, A., Bicsak, E., Karadi, K., Manning, J. & Kallai, J. (2003). 'Sex role identity related to the ratio of second to fourth digit length in women'. *Biological Psychology* 62: 147–56.

**10** Manning, J. T. (2002). *Digit Ratio: A Pointer to Fertility, Behaviour, and Health*. New Brunswick, NJ: Rutgers University Press. 68–71.

**11** Saad, G., Nepomuceno, M. V. & Mendenhall, Z. (2011). 'Testosterone and domain-specific risk: Digit ratios (2D:4D and rel2) as predictors of recreational, financial, and social risk-taking behaviours'. *Personality and Individual Differences* 51 (4).

**12** Kuhnen, C. M. & Chiao, J. Y. (2009). 'Genetic determinants of financial risk taking'. *PLoS ONE* 4: e4362.

**13** Interview on BBC World Service, 'Discovery' programme, May 2012.

**14** Kosfeld, M., Heinrichs, M., Zak, P. J., Fischbacher, U. & Fehr, E. (2005). 'Oxytocin increases trust in humans'. *Nature* 435: 673–6.

**15** BBC World Service interview op. cit.

## Chapter 8 : The Love Impulse –
## 'It Only Takes a Moment'

**1** Russell, B. (1969). *The Autobiography of Bertrand Russell, 1872–1914* Vol. 1. London: George Allen and Unwin Ltd. 75–82.

**2** Ibid. 75.

**3** Desmond, A. & Moore, E. (1991). *Darwin*. London: Michael Joseph. 257.

**4** Martineau, H. (1983). *Autobiography* Vol. 2. London: Virago. 175–7.

**5** Joyce, J. (1922/2010). Ulysses. Ware, Hertfordshire: Wordsworth Classics. 682.

**6** Liebowitz, M. R. (1983). *The Chemistry of Love.* New York: Little Brown. 37–49.

**7** Izard, C. E. (1960). 'Personality similarity and friendship'. *Journal of Abnormal Social Psychology* 67: 404–08.

**8** Lewis, D. (1985) op. Cit. 15–18.

**9** Levinger, G. & Breedlove, J. (1969). 'Interpersonal attraction and agreement: A study of marriage partners'. *Journal of Personality and Social Psychology* 3: 367–72.

**10** Furnham, A., Baguma, P. (1994). 'Cross-cultural differences in the evaluation of male and female body shapes'. *International Journal of Eating Disorders* 15 (1): 81–9.

**11** Photographer Julian Wolkenstein has developed a technique in which he creates perfectly symmetrical faces. He first takes a traditional, passport-style, full-face photograph then slices it down the centre. Next he flips one side horizontally and places it against the same side to create two separate portraits of the right side and left side of the face. See http://ifitshipitshere.blogspot.com/2011/02/echoism-your-left-side-vs-your-right.html

**12** Ford, C. S. & Beach, F. A. (1951). *Patterns of Sexual Behaviour.* New York: Harper & Row. 90–94.

**13** Jones, B. C., Little, A. C., Burt, D. M. & Perrett, D. I. (2004). 'When facial attractiveness is only skin deep'. *Perception* 33 (5): 569–76.

**14** Møller, A. P., Soler, M. & Thornhill, R. (1995). 'Breast asymmetry, sexual selection and human reproductive success'. *Ethology and Sociobiology* 16 (3): 207–19.

**15** Furnham & Baguma (1994) op. cit.

**16** Lavrakas, P. J. (1975). 'Female preferences for male physiques'.

Paper presented at the meeting of the Midwestern Psychological Association, Chicago, May.

17 Rose, C. (2011). 'The relevance of Darwinian selection to an understanding of visual art'. Personal communication.

18 Marlowe, F. & Wetsman, A. (1999). 'Preferred waist-to-hip ratio and ecology'. *Personality and Individual Differences* 30 (3): 481–9.

19 Singh, D. (1993). 'Adaptive significance of female physical attractiveness: Role of waist-to-hip ratio'. *Journal of Personality and Social Psychology* 65 (2): 293–307.

20 Until puberty both boys and girls have a near identical WHR at around 0.9. From that point on, however, the influence of oestrogen causes the woman's pelvis to grow, while men remain unaffected. In males the waist-to-hip ratio is little changed after puberty, with an ideal WHR of 0.9.

21 Furnham, A., McClelland, A. & Omer, L. (2003). A cross-cultural comparison of ratings of perceived fecundity and sexual attractiveness as a function of body weight and waist-to-hip ratio'. *Psychology, Health and Medicine* 8 (2): 219–30.

22 Mazur, A., (1986) 'U.S. trends in feminine beauty and overadaptation'. *Journal of Sex Research* 22: 281–303.

23 Marlowe & Wetsman (2001) op. cit.

24 The first utensils for drinking tea were imported from China. These lacked handles and were called 'tea bowls'. Not until around 1750 did the first cups with handles, the invention of one Robert Adams, arrive on the scene.

25 Sinclair, R., Hoffman, C., Mark, M., Martin, L. & Pickering, T. (1994). 'Construct accessibility and the misattribution of arousal: Schachter and Singer revisited'. *Psychological Science* 5 (1): 15–19.

**26** Meston, C. M. & Frohlich, P. F. (2003). 'Love at first fright: Partner salience moderates roller-coaster-induced excitation transfer'. *Archives of Sexual Behavior.* 32 (6): 537–44.

**27** Dutton, D. & Aron, A. (1974). 'Some evidence for heightened sexual attraction under conditions of high anxiety'. *Journal of Personality and Social Psychology* 30 (4): 510–17.

**28** Cohen, B., Waugh, G. & Place, K. (1989). 'At the movies: An unobtrusive study of arousal-attraction'. *Journal of Social Psychology* 129 (5): 691–3.

**29** Lavater, J. C. (1880). *Essays on Physiognomy; for the Promotion of the Knowledge and the Love of Mankind.* Gale Document Number CW114125313. Retrieved 15 May 2005 from Gale Group, Eighteenth Century Collections Online. (Original work published 1772.)

**30** Zebrowitz, L. A. (1999). *Reading Faces: Window to the soul?* Boulder, CO: Westview Press. 116–139.

**31** Wiggins, J. S., Wiggins, N. & Conger, J. C. (1968). 'Correlates of heterosexual somatic preference'. *Journal of Personality and Social Psychology* 10 (1): 82–90.

**32** Beck, S. B., Ward-Hull, C. I. & McLear, P. M. (1976). 'Variables related to women's somatic preferences of the male and female body'. *Journal of Personality and Social Psychology* 34 (6): 1200–10.

**33** Coward, R. (1984). *Female Desire.* London: Paladin. 231.

**34** In 2001 *Forbes* magazine estimated the total value of porn in the USA to be between $2.6 and $3.9 billion, including video, internet, magazines and pay-per-view.

**35** Ellis, B. J. (1992). 'The evolution of sexual attraction: Evaluative mechanisms in women'. In Barkow, J., Cosmides, L. & Tooby, J.

(eds.) *The Adapted Mind*. New York: Oxford University Press. 267–88.

**36** Pennebaker, J. W., Dyer, M. A., Caulkins, R. S., Litowitz, D. L., Ackreman, P. L., Anderson, D. B. & McGraw, K. M. (1979). 'Don't the girls get prettier at closing time: A country and western application to psychology'. *Personality and Social Psychology Bulletin* 5 (1): 122.

**37** Conley, T. D., Moors, A. C., Matsick, J. L., Ziegler, A. & Valentine, B.A. (2011). 'Women, men, and the bedroom: Methodological and conceptual insights that narrow, reframe, and eliminate gender differences in sexuality'. *Current Directions in Psychological Science* 20 (5): 296–300.

**38** Petersen, J. L. & Hyde, J. S. (2010). 'A meta-analytic review of research on gender differences in sexuality, 1993–2007'. *Psychological Bulletin* 136: 21–38.

**39** Clark, R. D. & Hatfield, E. (1989). 'Gender differences in receptivity to sexual offers'. *Journal of Psychology & Human Sexuality* 2: 39–45.

**40** Conley et al. (2011) op. cit.

**41** Finkel, E. J. & Eastwick, P. W. (2009). 'Arbitrary social norms influence sex differences in romantic selectivity'. *Psychological Science* 20: 1290–5.

**42** Fisher, T. D., Moore, Z. T. & Pittenger, M. J. (2011). 'Sex on the brain? An examination of frequency of sexual cognitions as a function of gender, erotophilia, and social desirability'. *Journal of Sex Research* 49 (1): 69–77.

**43** Conley et al. (2011) op. cit.

## Chapter 9: The Overeating Impulse – Digging Our Graves with Our Teeth

1 Wansink, B. (2006). *Mindless Eating*. London: Bantam Books. 106–23.

2 The term 'obese' has been criticised in some UK government circles as derogatory and the suggestion made that 'it might be better to refer to a "healthier weight" rather than "obesity" – and to talk more generally about health and wellbeing or specific community issues.' This advice was, ironically, printed in a paper with the title *Obesity: Working With Local Communities*. Tam Fry, spokesman for the National Obesity Forum, commented: 'They should be talking to people in an adult fashion. There should be no problem with using the proper terminology. If you beat around the bush then you confuse people. This is extremely patronising. Obesity is a well-defined, World Health Organisation standard that everybody can understand.'

3 The BMI is calculated by multiplying the person's weight in pounds by 703, and then dividing by the square of their height in inches. For example, someone who weighed 150 lbs and was 5' 8" tall would have a BMI of 22. 8 (i.e. $150 \times 703/68^2$ ) and be within the ideal weight range. A weight gain of a stone (14 lbs) would just place them in the overweight category (BMI = 25), while if their weight ever soared to 199 lbs then they would be obese (BMI = 30).

4 Devised more than two centuries ago by Lambert Adolphe Jacques Quetelet, the BMI when applied to individuals as opposed to populations is pseudostatistics masquerading as scientific fact. Quetelet was a brilliant Belgian mathematician, not a medical doctor, who devised his formula at the request of his government. They wanted a quick and easy way to measure the extent of obesity among the Belgian population so as to better allocate funds. He specifically stated it should not, indeed could not, be used as a

measure of individual fatness. Here are six good reasons for not trusting the BMI:

1. The BMI makes no sense scientifically. Quetelet was obliged to square the individual's height in order to produce a formula in line with the overall data.

2. The formula fails to take into account the relative proportions of bone, muscle and fat. Since bone is denser than muscle and twice as dense as fat, a combination of low fat with strong bones and good muscle tone will produce a high BMI. By this reasoning many world-class athletes are overweight or even obese.

3. While the formula works reasonably well for sedentary people who combine a high relative fat content with low muscle mass, it gives entirely the wrong answer in the case of those who are fit, healthy and lean.

4. Quetelet is the mathematician who dreamed up the idea of 'the average man'. That's fine when it relates to an entire population but absurd if applied to an individual. It leads to nonsensical statements such as all of us have 1.5 legs (some people have only one leg but no one has three or more!) and 2.4 children.

5. The BMI has been described as mathematical snake oil. The single, percentage-like number it provides carries an air of undeserved scientific authority.

6. It incorrectly and nonsensically suggests that there are four separate weight categories – under, over, obese and just right – whose precise boundaries depend on a decimal place.

5 Wang, Y.C., McPhersin, K., Marsh, T., Gortmaker, S.L. & Brown, M. (2011). 'Health and economic burden of the projected obesity trends in the USA and UK', *Lancet* 378 (9793): 815–25.

**6** Chester, J. & Montgomery, K. (2007). 'Interactive food & beverage marketing: Targeting children and youth in the digital age'. Report from Berkeley Media Studies Group, May.

**7** Berridge, K. (1996). 'Food reward: Brain substrates of wanting and liking'. *Neuroscience and Biobehavioural Reviews* 20: 1–25.

**8** Ventura, A. K. & Mennella, J. A. (2011). 'Innate and learned preferences for sweet taste during childhood'. *Current Opinions in Clinical Nutritional and Metabolic Care* 14 (4): 379–84.

**9** Berthoud, H. R. (2011). 'Metabolic and hedonic drives in the neural control of appetite: Who is the boss?'. *Current Opinions in Clinical Nutritional and Metabolic Care* 21 (6): 888–96.

**10** Coll, A. P., Farooqi, I. S. & O'Rahilly, S. (2007). 'The hormonal control of food intake'. *Cell* 129 (2): 251–62.

**11** Cota, D., Tschop, M. H., Horvath, T. L. & Levine, A. S. (2006). 'Cannabinoids, opioids and eating behavior: The molecular face of hedonism?'. *Brain Research Review* 51 (1): 85–107.

**12** Erlanson-Albertsson, C. (2005). 'Sugar triggers our reward-system. Sweets release opiates which stimulates the appetite for sucrose – insulin can depress it'. *Lakartidningen* 192 (21): 1620–2, 1625, 1627.

**13** Kelley, A. E., Bakshi, V. P., Haber, S. N., Steininger, T. L., Will, M. J. & Zhang, M. (2002). 'Opioid modulation of taste hedonics within the ventral striatum'. *Physiological Behaviour* 76 (3): 365–77.

**14** Bisogni, C. A., Falk, L. W., Madore, E., Blake, C. E., Jastran, M., Sobal, J. et al. (2007). 'Dimensions of everyday eating and drinking episodes'. *Appetite* 48 (2): 218–31.

**15** Beaver, J. D., Lawrence, A. D., van Ditzhuijzen, J., Davis, M.

H., Woods, A., & Calder, A. J. (2006). 'Individual differences in reward drive predict neural responses to images of food'. *Journal of Neuroscience* 26 (19): 5160–6.

**16** Wang, G. J., Volkow, N. D., Logan, J., Pappas, N. R., Wong, C. T., Zhu, W. et al. (2001). 'Brain dopamine and obesity'. *Lancet* 357 (9253): 354–7.

**17** Nederkoorn, C., Jansen, E., Mulkens, S. & Jansen, A. (2006). 'Impulsivity predicts treatment outcome in obese children'. *Behaviour Research and Therapy.* 45 (5): 1071–5.

**18** Pelchat, M., Johnson, A., Chan, R., Valdex, J. & Ragland, J. D. (2004). 'Images of desire: Food-craving activation during fMRI'. *Neuroimage* 23: 1486–93.

**19** Schlosser, E. (2001). *Fast Food Nation.* New York: Houghton Mifflin Co. 3.

**20** Wang et al. (2001) op. cit.

**21** Johnson, P. M. & Kenny, P. J. (2010). 'Dopamine D2 receptors in addiction-like reward dysfunction and compulsive eating in obese rats'. *Nature Neuroscience* 13: 635–41.

**22** Ackerman, J. (2012). 'The ultimate social network'. *Scientific American* 306 (6) July: 21–7.

**23** Blaser, M. J. (2005). 'Global warming and the human stomach: Microecology follows macroecology'. *Transactions of the American Clinical and Climatological Association* 116: 65–76.

**24** Maddock, J. (2004). 'The relationship between obesity and the prevalence of fast food restaurants: state-level analysis'. *American Journal of Health Promotion* (19): 137–43.

**25** Wang, G. J., Volkow, N. D., Thanos, P. K. & Fowler, J. S. (2004). 'Similarity between obesity and drug addiction as assessed by neurofunctional imaging: a concept review'. *Journal of Addictive Disease* 23 (3): 39–53.

26 'Fresh fruit, hold the insulin', *Scientific American* 306 (5): 7. Comment. Board of Editors.

27 Ibid.

28 'McDonald's orders up augmented reality from total immersion, in global promotion for Fox's "Avatar"'. Serious Games Market, 17 December 2009, http://seriousgamesmarket.blogspot.com/2009/12/serious-games-as-ar-extensive.html (viewed 10 April 2010).

29 Wansink, B. & van Ittersum, K. (2006). 'Ice cream illusions. Bowls, spoons, and self-served portion sizes'. *American Journal of Preventive Medicine* 31 (3): 240–3.

30 Wansink, B. & van Ittersum,K. (2003). 'Bottoms up! The influence of elongation on pouring and consumption volume'. *Journal of Consumer Research* 30 (December): 455–63.

31 Wansink, B. (2010) *Mindless Eating. Why We Eat More Than We Think.* London: Bantam Books. 189.

32 Benedict, C., Brooks, S. J., O'Daly, O. G., Almèn, M. S., Morell, A., Åberg, K., Gingnell, M., Schultes, B., Hallschmid, M., Broman, J.-E., Larsson, E.-M. & Schiöth, H. B. 'Acute sleep deprivation enhances the brain's response to hedonic food stimuli: an fMRI study'. *Journal of Clinical Endocrinology and Metabolism.* Publishing online 97/3/E443.

33 Naek, D.T., Wood, W., Wu, M. & Kurlander, D. (2012). 'The pull of the past: When do habits persist despite conflict with motives?'. *Personality and Social Psychology Bulletin* 37 (11): 1428–37.

# Chapter 10: The Buying Impulse – The How and Why of What We Buy

1 Sorensen, H. (2009). *Inside the Mind of the Shopper*. New Jersey: Wharton School Publishing. 8.

2 Applebaum, W. (1951). 'Studying customer behaviour in retail stores'. *Journal of Marketing*, 16 (October): 172–8.

3 Stern, H. (1962). 'The significance of impulse buying today'. *Journal of Marketing* (April): 59–60.

4 Rook, D. W. (1987). 'The buying impulse'. *Journal of Consumer Research* 14 (September): 191.

5 Hausman, A. (2000). 'A multi-method investigation of consumer motivations in impulse buying behavior'. *Journal of Consumer Marketing* 17 (5): 403–17.

6 Rook (1987) op. cit.

7 I use the term supermarket generically despite the fact they come in all sizes from relative minnows to hangar-like giants with 100,000 or more square feet of retail space.

8 Blythman, J. (2004). *Shopped*. London: Fourth Estate. 15.

9 There has long been a debate about who opened the world's first true supermarket. To bring an end to the arguments, the US Food Marketing Institute and the Smithsonian Institution researched the issue and concluded that Michael J. Cullen can rightly claim the distinction.

10 Lebhar, G. M. (1963). *Chain Stores in America, 1859–1962*. 3rd edn. New York: Chain Store Publishing Corporation. 226–8.

11 Levitt, T. (1975). 'Marketing myopia'. *Harvard Business Review*. September–October. No. 75507.

**12** Scamell-Katz, S. (2012) *The Art of Shopping. How We Shop and Why We Buy.* London: LID Publishing. 69.

**13** Quoted from: Derbyshire, D. (2004) 'They have ways of making you spend'. *Daily Telegraph.* 31 December.

**14** Scamell-Katz (2012) op. cit. 111.

**15** Derbyshire (2004) op. cit.

**16** James, W. (1890). *The Principles of Psychology.* New York: Holt. 291–2.

**17** Ways in which the shopping experience can be enhanced as described in: Lewis, D. & Bridger, D. (2003). *The Soul of the New Consumer: Authenticity, What We Buy and Why in the New Economy.* London: Nicholas Brealey. 128–47.

**18** Underhill, P. (1999). *Why We Buy. The Science of Shopping.* New York: Simon & Schuster. 11.

## Chapter 11: The Imitation Impulse –
## 'A Beautiful Place to Die'

**1** Cooley, C. H. (1998). *On Self and Social Organization.* Chicago: University of Chicago Press. 20–2.

**2** I have developed a measurement scale, the milli-Helen, to assess potential role models. It is derived from the mythical Helen of Troy, whose abduction by Paris led to the Trojan Wars. Of her beauty the dramatist Christopher Marlowe wrote: 'Was this the face that launch'd a thousand ships, And burnt the topless towers of Ilium?' A milli-Helen is the celebrity's ability to launch a single ship. Or in modern commercial terms, to persuade only a few people to copy them. The higher the milli-Helen score the greater that celebrity's commercial value as a brand ambassador.

**3** Since its construction, in 1937, more than 1,500 people have ended their lives by leaping into the ocean from the 245ft-high deck of San Francisco's Golden Gate Bridge. Indeed suicides and attempted suicides are so frequent that, in 2006, documentary filmmaker Eric Steel set up a camera to record them. During the few months of the filming, 24 people jumped from the bridge, approximately one every 15 days. Source: Matier, P. & Ross, A. (2005). 'Film captures suicides on Golden Gate Bridge'. *San Francisco Chronicle*, 19 January.

**4** Interview with Sussex coastguard Don Ellis, 23 October 2011.

**5** Ibid.

**6** Ibid.

**7** *Daily Telegraph* 4 August 2011.

**8** My use of the terms 'riot', 'rioters', 'looting' and 'looters' is based on common parlance and should not be taken as indicating my agreement with these descriptors or with my adoption of any particular political position.

**9** Malkani, G. (2011). 'Britain burns the colour of "A Clockwork Orange"'. *Financial Times* 13/14 August. 9.

**10** Reicher, S. & Stott, C. (2012). *Mad Mobs and Englishmen? Myths and Realities of the 2011 Riots.* Kindle edition.

**11** Ibid.

**12** Lambert, O. (2012). 'My child the rioter'. *Wonderland.* BBC 2, 31 January.

**13** A review of the ways in which people ended their lives in 56 countries identified hanging as the most frequent method, accounting for just over half of male and one in four female suicides. In the USA, just over half of all self-inflicted deaths involve the use of firearms, although asphyxiation and poisoning are also fairly common and together make up about 40 per cent of that country's suicides. Other methods include slitting the throat or wrist, non-accidental drowning, setting oneself ablaze, intentional starvation, electrocution, overdosing on pills and taking poisons. Deaths from blunt-force trauma include leaping from high buildings, bridges or cliff tops, throwing oneself through a high window, jumping under a train, bus or truck and deliberately crashing a car, motorbike, speedboat or aircraft. Source: Ajdacic-Gross, V. et al. (2008). 'Methods of suicide: International suicide patterns derived from the WHO mortality database'. *Bulletin of the World Health Organisation* 86 (9): 726–32.

14 Stack, S. (2003). 'Media coverage as a risk factor in suicide'. *Journal of Epidemiology & Community Health* 57 (4): 238–40.

15 Suicide, from the Latin *sui caedere* ('to kill oneself'), is the tenth leading cause of death worldwide with rates having risen sharply, especially in Ireland and Greece, during the years of austerity. They are now 60 per cent higher than half a century ago, with most of this increase occurring in the industrialised nations. Around a million people take their own lives each year and between 10 and 20 million attempt to do so. In the USA, suicides outnumber homicides by nearly two to one, making it the 11th leading cause of death, ahead of liver disease and Parkinson's disease. Although women attempt suicide more often than men, males are more successful. Some experts believe this is because men use more violent and effective means, such as guns or nooses, while women employ less severe methods, for example by swallowing an overdose. Drug overdoses account for about two-thirds of suicides among women and one-third among men. Sources: 'Suicide Prevention'. *WHO Sites: Mental Health.* 16 February 2006.

http://www.who.int/mental_health/prevention/suicide/suicidepre-vent/en/. 'USA Suicide: 2007 Official Final Data'. *Association of Suicidology*. www.suicidology.org.

**16** Becker, K., Schmidt, M. H. (2005). 'When kids seek help on-line: Internet chat rooms and suicide'. *Reclaiming Children and Youth: The Journal of Strength-Based Interventions* 13 (4): 229.

**17** Baker, F. (2009). 'Inquest rules on gay teen goaded to suicide by baying Derby mob'. *Pink News* 16 January.

**18** Quoted from Louis de Bernières (1996) 'Legends of the Fall'. *Harper's Magazine*. January.

**19** Ellis (2011) op. cit.

**20** Ramachandran, V. S. (2000). 'Mirror neurons and imitation learning as the driving force behind "the great leap forward" in human evolution'. *Edge*. http://www.edge.org/3rd_culture/ramachandran/ramachandran_p1.html

**21** Ibid.

**22** Iacoboni, M., Woods, R. P., Brass, M., Bekkering, H., Mazziotta, J. C. & Rizzolatti, G. (1999). 'Cortical mechanisms of human imitation'. *Science* 286: 2526–8.

**23** Rizzolatti, G. & Arbib, M. A. (1998). 'Language within our grasp'. *Trends in Neurosciences* 21: 188–94.

**24** Le Bon, G. (1895/2002). *The Crowd. A Study of the Popular Mind*. New York: Dover Publications. 8.

**25** Reicher and Stott (2012) op. cit.

**26** Schachter, S. & Singer, J. (1962). 'Cognitive, social, and physiological determinants of emotional state'. *Psychological Review* 69: 379–99.

**27** Reicher and Stott (2012) op.cit.

# Chapter 12: Deplete Us Not Into Temptation

1 Guerber, H.A. (1927). *Myths of Greece and Rome.* London: George G. Harrap & Co. 313.

2 Hofmann, W., Baumeister, R. F., Förster, G. & Vohs, K. D. (2011). 'Everyday temptations: An experience sampling study of desire, conflict, and self-control'. *Journal of Personality and Social Psychology.* 102 (6): 1318–35.

3 Baumeister, R. F. (2008). 'Free will in scientific psychology'. *Perspectives on Psychological Science* 3 (1): 14–19.

4 Mischel, H. N. & Mischel, W. (1983). 'The development of children's knowledge of self-control strategies'. *Child Development* 54: 603–19.

5 McClure, R. F. (1986). 'Self control and achievement motivation in young and old subjects'. *Psychology: A Journal of Human Behaviour* 23 (1): 20–2.

6 Duckworth, A. L. & and Seligman, M. E. P. (2005) 'Self-discipline outdoes IQ in predicting academic performance of adolescents'. *Psychological Science* 16 (12): 939–44.

7 Baumeister, R. F. (2012) 'Self-control – the moral muscle'. *The Psychologist* 25 (2): 112–15.

8 Vohs, K. D. & Schooler, J. W. (2008). 'The value of believing in free will: Encouraging a belief in determinism increases cheating'. *Psychological Science* 19: 49–54.

9 Baumeister, R. F., Vohs, K. D. & Tice, D. M. (2007). 'The strength model of self-control'. *Current Directions in Psychological Science* 16 (6): 351–5.

10 Nordgren, L. F. & Chou, E. Y. (2011). 'The push and pull of temptation: The bidirectional influence of temptation on self-control'. *Psychological Science* 22 (11): 1386–90.

**11** Baumeister, R. F., Bratslavsky, E., Muracen, M. & Tice, D. M. (1998). 'Ego depletion: Is the active self a limited resource?'. *Journal of Personality and Social Psychology* 74: 1252–6.

**12** Muraven, M., Lorraine, C. R. & Neinhaus, K. (2002). 'Self-control and alcohol restraint: An initial application of the self-control strength model'. *Psychology of Addictive Behaviours* 16 (2) June: 113–20.

**13** Baumeister, R. F., Sparks, E. A., Stillman, T. F. & Vohs, K. D. (2008). 'Free will in consumer behavior: Self-control, ego depletion, and choice'. *Journal of Consumer Psychology* 18: 4–13.

**14** Loewenstein, G. (1996). 'Emotions in economic theory and economic behaviour'. *Preferences, Behaviour, and Welfare* 90 (2): 426–32.

**15** Vohs, K. D. & Faber, R. J. (2007). 'Spent resources: Self-regulatory resource availability affects impulse buying'. *Journal of Consumer Research* 33: 537–47.

**16** Wargo, E. (2009). 'Resisting temptation'. *Observer* 22.

**17** Vohs, K. D. & Heatherton, T. F. (2000). 'Self-regulatory failure: A resource depletion approach.' *Psychological Science* 11: 249–54.

**18** Inzlicht, M., McKay, L., & Aronson, J. (2006). 'Stigma as ego depletion: How being the target of prejudice affects self-control'. *Psychological Science* 17: 262–9.

**19** Baumeister (2012) op. cit.

**20** Inzlicht, M. & Gutsell, J. N. (2007). 'Running on empty – neural signals for self-control failure'. *Psychological Science* 18 (11): 933–7.

**21** Gailliot, M. T., Baumeister, R. F., DeWall, C. N., Maner, J. K., Plant, E. A., Tice, D. M. et al. (2007). 'Self-control relies on glucose as a limited energy source: Willpower is more than a metaphor'. *Journal of Personality and Social Psychology* 92: 326.

**22** Kurzban, R. (2010). 'Does the brain consume additional glucose during self-control tasks?'. *Evolutionary Psychology* 8 (2): 246.

**23** Comfort, A. (1967). *The Anxiety Makers: Some Curious Preoccupations of the Medical Profession.* London: Nelson: 70–113.

**24** In fact, the sin of Onan, for which he was struck dead by God, was neither masturbation nor even coitus interruptus but refusing to father children for his deceased brother (Genesis 38:9).

**25** Comfort (1967) op. cit.

**26** Ainslie, G. (1975). 'Specious reward: A behavioral theory of impulsiveness and impulse control'. *Psychological Bulletin* 82: 463.

**27** Baumeister, R. F., Gailliot, M., DeWall, C. N. & Oaten, M. (2006). 'Self-regulation and personality: How interventions increase regulatory success, and how depletion moderates the effects of traits on behaviour'. *Journal of Personality* 74 (6) December: 1743–802.

**28** Solomon, S., Greenberg, J., & Pyszczynski, T. (2004). 'Lethal consumption: Death-denying materialism.' in T. Kasser & A. Kanner (Eds.). *Psychology and consumer culture: The struggle for a good life in a materialistic world.* Washington, DC: American Psychological Association. 131–2.

**29** Culotta, E. (2009). 'On the origin of religion'. *Science* 326: 784–787.

**30** Kay, A. C., Moscovitch, D. A. & Laurin, K. (2010). 'Randomness, attributions of arousal, and belief in God'. *Psychological Science* 21: 216–18.

**31** Roes, F. L. & Raymond, M. (2003). 'Belief in moralizing gods'. *Evolution & Human Behavior* 24: 126–35.

**32** Rounding, K., Lee, A., Jacobson, J.A. & Ji, L.-J. (2012). 'Religion replenishes self-control'. *Psychological Science* 23 (6): 635–42.

## Afterword: Free Will Is a Grand Illusion

**1** Burns, J. M. & Swerdlow, R. H. (2003). 'Right orbitofrontal tumor with pedophilia symptom and constructional apraxia sign'. *Archives of Neurology* 60: 437–40.

**2** Morse, S. (2006). 'Brain overclaim syndrome and criminal responsibility: a diagnostic note'. *Ohio State Journal of Criminal Law* 3: 397–412.

**3** Raine, A. (1993). *The Psychopathology of Crime: Criminal Behaviour as a Clinical Disorder.* San Diego: Academic Press. 377.

**4** Golding, S. L., Roesch, R. & Schreiber, J. (1984). 'Assessment and conceptualization of competency to stand trial: preliminary data on the Interdisciplinary Fitness Interview'. *Law and Human Behaviour.* 8 (3/4): 321–34.

**5** Fazel, S. & Danesh, J. (2002). 'Serious mental disorder in 23000 prisoners: a systematic review of 62 surveys'. *Lancet* 359: 545–50.

**6** Widom, C. S. (1989). 'The cycle of violence'. *Science* 244: 160–6.

**7** Baumeister, R. F. (2012). 'Self-control – the moral muscle'. *The Psychologist* 25 (2): 112–15.

**8** Wegner, D.M. (2002). 'Free Will in Scientific Psychology'. *Perspectives on Psychological Science* 3 (1): 14–19, cited from page 14.

**9** Baumeister, R. F. (2008). 'Free will in scientific psychology'. *Perspectives on Psychological Science* 3 (1): 14–19.

**10** Hallett, M. (2007). 'Volitional control of movement: The physiology of free will'. *Clinical Neurophysiology* 118: 1179–92.

**11** Libet, B., Gleason, C. A., Wright, E. W., & Pearl, D. K. (1983). Time of conscious intention to act in relation to onset of cerebral activity (readiness potential): The unconscious initiation of a freely voluntary act'. *Brain* 106: 623–42.

**12** Soon, C. S., Brass, M., Heinze, H.-J., & Haynes, J.-D. (2008). 'Unconscious determinants of free decisions in the human brain'. *Nature Neuroscience* 11: 543–5.

**13** Minsky, M. (1986). *The Society of Mind.* New York: Simon and Schuster. 339.

**14** Blackmore, S. (2007). 'Mind over matter?'. http://www.patheos. com/blogs/monkeymind/2007/08/susan-blackmore-on-ben-libet. html.

**15** Libet et al. (1983) op. cit.

**16** This topic is extensively and brilliantly addressed in Daniel M.Wegner's book *The Illusion of Conscious Will* (2002) Massachusetts Cambridge: MIT Press. The first two chapters, pages 1–61 offer an insightful introduction to this important and controversial topic.

**17** Blackmore (2007) op. cit.

**18** Lau, H. C., Rogers, R. D. & Passingham, R. E. (2006). 'On measuring the perceived onsets of spontaneous actions'. *Journal of Neuroscience* 26: 7265–71.

**19** Lau, H. C., Rogers, R. D. & Passingham, R. E. (2007). 'Manipulating the experienced onset of intention after action execution'. *Journal of Cognitive Neuroscience* 19: 81–90.

**20** Ajzen, I. (2002). 'Perceived behavioral control, self-efficacy, locus of control, and the theory of planned behavior'. *Journal of Applied Social Psychology* 32: 665–83.

**21** Bandura, A. (1982). 'Self-efficacy in human agency'. *American Psychologist* 37: 122–147.

**22** Rigoni, D., Kuhn, S., Sartori, G. & Brass, M. (2011). 'Inducing Disbelief in free will alters brain correlates of preconscious motor preparation: The brain minds whether we believe in free will or not'. *Psychological Science* 22 (5):613–18.

**23** Vohs, K. D. & Schooler, J. W. (2008). 'The value of believing in free will: Encouraging a belief in determinism increases cheating'. *Psychological Science* 19: 49–54.

**24** Rigoni et al. (2011) op. cit.

**25** Baumeister, R. F., Masicampo, E. J., & DeWall, C. N. (2009). 'Prosocial benefits of feeling free: Disbelief in free will increases aggression and reduces helpfulness'. *Personality and Social Psychology Bulletin* 35: 260–8.

**26** Vohs & Schooler (2008) op. cit.

**27** Solomon, S., Greenberg, J., & Pyszcynski, T. (2004). 'Lethal consumption: Death-defying materialism'. In Kasser ,T. & Kanner, A. D. (eds.) *Psychology and Consumer Culture*. Washington D.C.: American Psychological Association. 131.

**28** Ibid.

**29** Wegner, D. (2004) 'Précis of *The Illusion of Conscious Will*'. *Behavioural and Brain Sciences* 27 (5) 652–59.

# Bibliography

Ackerman, J. M., Goldstein, N. J., Shapiro, J. R. & Bargh, J. A. (2009). 'You Wear Me Out: The Vicarious Depletion of Self-Control'. *Psychological Science* 20(3): 326–32.

Adamson, R. E. (1952). 'Functional Fixedness as Related to Problem Solving: a Repetition of Three Experiments'. *Journal of Experimental Psychology* 44: 288–91.

Ahmadlou, M. & Adeli, H. (2011). 'Functional Community Analysis of Brain: A New Approach for EEG-Based Investigation of the Brain Pathology'. *Neuroimage* 58: 401–8.

Ainslie, G. (1975). 'Specious Reward: A Behavioral Theory of Impulsiveness and Impulse Control'. *Psychological Bulletin* 82: 463–96.

Ajdacic-Gross, V. Et al. (2008). 'Methods of Suicide: International Suicide Patterns Derived from the WHO Mortality Database'. *Bulletin of the World Health Organisation* 86(9): 726–32.

Alderman, N. (2011). 'Effectiveness of Neurobehavioural Rehabilitation for Young People and Adults with Traumatic Brain Injury and Challenging Behaviour'. *ACNR,* 11(4): 26–7.

Aldhous, P. (1998). 'What Chance We've Got the Wrong Man?' *New Scientist* No. 2123, 28th February: 20.

Ambady, N. & Gray, H. M. (2002). 'On Being Sad and Mistaken: Mood Effects on the Accuracy of Thin-Slice Judgments'. *Journal of Personality and Social Psychology* 83(4): 947–61.

Anderson, A. K., Christoff, K., Stappen, I., Panitz, D., Ghahremani, D. G., Glover, G., Gabrieli, J. D. E. & Sobel, N. (2003). 'Dissociated Neural Representations of Intensity and Valence in Human Olfaction'. *Nature Neuroscience* 6(2): 196–202.

Anonymous. (1851). 'A Most Remarkable Case'. *American Phrenological Journal and Repository of Science, Literature, and General Intelligence* 13, 89: col 3.

Applebaum, W. (1951). 'Studying Customer Behavior in Retail Stores'. *The Journal of Marketing* 16 (October): 172–8.

Ariely, D. (2008). *Predictably Irrational: The Hidden Forces that Shape our Decisions.* New York: Harper.

Asaad, W. F. & Eskandar, E. N. (2011). 'Encoding of Both Positive and Negative Reward Prediction Errors by Neurons of the Primate Lateral Prefrontal Cortex and Caudate Nucleus'. *The Journal of Neuroscience* 31(49): 17772–87.

Astrup, A. (2008). 'Dietary Management of Obesity'. *JPEN Journal of Parenteral Enteral Nutrition* 32(5): 575–77.

Atkinson, A. P., Thomas, M. & Cleeremans, A. (2000). 'Consciousness: Mapping the Theoretical Landscape'. *Trends in Cognitive Sciences* 4: 372–82.

Austin, E. J., Farrelly, D., Black, C. & Moore, H. (2007). 'Emotional Intelligence, Machiavellianism and Emotional Manipulation: Does EI Have a Dark Side?' *Personality and Individual Differences* 43(1): 179–89.

Bailey, A. A. & Hurd, P. L. (2005). 'Finger Length Ratio (2D:4D) Correlates with Physical Aggression in Men but not in Women'. *Biological Psychology* 68 (3): 215–22.

Bainbridge, D. (2009). *Teenagers: A Natural History.* London: Portobello Books.

Baker, F. (2009). *Inquest* 'Rules On Gay Teen Goaded To Suicide By Baying Derby Mob'. *Pink News* 16th January.

Bancroft, M. D., Janssen, E., Strong, D., Carnes, L., Vukadinovic, Z. & Long, J. S. (2003). 'Sexual Risk-Taking in Gay Men: The Relevance of Sexual Arousability, Mood, and Sensation Seeking'. *Archives of Sexual Behaviour* 32(6): 555–72.

Bargh, J. A., Chen, M. & Burrows, L. (1996). 'Automaticity of Social Behavior: Direct Effects of Trait Construct and Stereotype Activation on Action'. *Journal of Personality and Social Psychology* 71(2): 230–44.

Bargh, J. A. & Chartrand, T. L. (1999). 'The Unbearable Automaticity of Being'. *American Psychologist* 54(7): 462–79.

Bargh, J. A. & Ferguson, M. J. (2000). 'Beyond Behaviorism: On the Automaticity of Higher Mental Processes'. *Psychological Bulletin* 126(6): 925–45.

Baron, R. A. (1981). 'Olfaction and Human Social Behaviour: Effects of a Pleasant Scent on Attraction and Social Perception'. *Basic and Applied Social Psychology* 1(2): 163–172.

Baron, R. A. (1981). 'Olfaction and Human Social Behaviour: Effects of a Pleasant Scent on Attraction and Social Perception'. *Personality and Social Psychology Bulletin* 7 (4): 611–16.

Barrett, L. F. (2006). 'Are Emotions Natural Kinds?' *Perspectives on Psychological Science* 1(1): 28.

Başar, E. (2010). *Brain-Body-Mind in the Nebulous Cartesian System: A Holistic Approach by Oscillations.* Springer Verlag.

Batterink, Y., Yokum, S. & Stice, E. (2010). 'Body Mass Correlates Inversely with Inhibitory Control in Response to Food among Adolescent Girls: An fMRI Study'. *Neuroimage* 52(4): 1696–703.

Baumeister, R. F. (2008). 'Free Will in Scientific Psychology'. *Perspectives on Psychological Science* 3(1): 14–19.

Baumeister, R. F. (2012). 'Self-Control – The Moral Muscle'. *The Psychologist* 25(2): 112–15.

Baumeister, R. F., Bratslavsky, E., Muraven, M. & Tice, D. M. (1998). 'Ego Depletion: Is the Active Self a Limited Resource?' *Journal of Personality and Social Psychology* 74(5): 1252–65.

Baumeister, R. F. & Vohs, K. D. (2007). 'Self-egulation, Ego Depletion, and Motivation'. *Social and Personality Psychology Compass* 1(1): 115–28.

Baumeister, R. F., Vohs, K. D. & Tice, D. M. (2007). 'The Strength Model of Self-Control'. *Current Directions in Psychological Science* 16 (6): 351–5.

Baumeister, R. F., Sparks, E. A., Stillman, T. F. & Vohs, K. D. (2008). 'Free Will in Consumer Behavior: Self-Control, Ego Depletion, and Choice'. *Journal of Consumer Psychology* 18: 4–13.

Beatty, S. E. & Ferrell M. E. (1998). 'Impulsive Buying: Modelling Its Precursors'. *Journal of Retailing,* 74(2): 169–91.

Beaver, J. D., Lawrence, A. D., Van Ditzhuijzen, J., Davis, M.H., Woods, A. & Calder, A. J. (2006). 'Individual Differences in Reward Drive Predict Neural Responses to Images of Food'. *Journal of Neuroscience* 26 (19): 5160–6.

Beaverbrook, M. A. (2010). *Success.* Kessinger Publishing.

Bechara, A., Damasio, H. & Damasio, A. R. (2000). 'Emotion, Decision Making and the Orbitofrontal Cortex'. *Cerebral Cortex* 10(3): 295–307.

Beck, S. B., Ward-Hull, C. I. & McLear, P. M. (1976). 'Variables Related to Women's Somatic Preferences of the Male and Female Body'. *Journal of Personality and Social Psychology* 34(6), 1200–10.

Becker, W. J., Cropanzano, R. & Sanfey, A. G. (2011). 'Organizational Neuroscience: Taking Organizational Theory Inside the Neural Black Box'. *Journal of Management,* 37(4), 933–61.

Beilock, S. L., Afremow, J. A., Rabe, A. L. & Carr, T. H. (2001). '"Don't Miss!" The Debilitating Effects of Suppressive Imagery of Golf Putting Performance'. *Journal of Sport & Exercise Psychology* 23(3): 200–21.

Benson, D. F. & Blumer, D. (Eds.) (1975). *Psychiatric Aspects of Neurologic Disease.* New York: Grune & Stratton Inc.

Benton, D. (1982). 'The Influence of Androstenol – a Putative Human Pheromone – on Mood Throughout the Menstrual Cycle'. *Biological Psychology* 15 (3–4): 249–56.

Berkman, E. T., Falk, E. B. & Lieberman, M. D. (2011). 'In the Trenches of Real-World Self-Control: Neural Correlates of Breaking the Link Between Craving and Smoking'. *Psychological Science* 22(4): 498–506.

Berlin, H. A., Rolls, E. T. & Kischka, U. (2004). 'Impulsivity, Time Perception, Emotion and Reinforcement Sensitivity in Patients with Orbitofrontal Cortex Lesions'. *Brain,* 127(5): 1108–26.

Bernstein, D. M. & Loftus, E. F. (2009). 'How to Tell if a Particular Memory is True or False'. *Perspectives on Psychological Science,* 4(4), 370–4.

Berridge, K. (1996) 'Food Reward: Brain Substrates of Wanting and Liking'. *Neuroscience and Biobehavioral Reviews* 20(1): 1–25.

Bigelow, H. J. (1850). 'Dr Harlow's Case of Recovery from the Passage of an Iron Bar through the Head'. *American Journal of the Medical Sciences* 20: 13–22. [Copy in Macmillan, M. (2000). *An Odd Kind of Fame. Stories of Phineas Gage.* Cambridge, Mass: The MIT Press].

Bijleveld, E., Custers, R. & Aarts, H. (2009). 'The Unconscious Eye Opener: Pupil Dilation Reveals Strategic Recruitment of Resources upon Presentation of Subliminal Reward Cues'. *Psychological Science* 20(11): 1313–15.

Biran, I. & Chatterjee, A. (2004). 'Alien Hand Syndrome'. *Archives of Neurology* 61: 292–4.

Bisogni, C. A., Falk, L. W., Madore, E., Blake, C. E., Jastran, M., Sobal, J. & Devine, C. M. (2007). 'Dimensions of Everyday Eating and Drinking Episodes'. *Appetite* 48(2): 218–31.

Blackburn, S. (2012). 'Mother Nature'. *New Statesman*, 16th January, 44.

Blais, A-R. & Weber, E. U. (2006). 'A Domain-Specific Risk-Taking (DOSPERT) Scale for Adult Populations'. *Judgment and Decision Making* 1(1): 33–47.

Block, J. (1995). 'A Contrarian View of the Five-Factor Approach to Personality Description'. *Psychological Bulletin* 117(2): 187–215.

Blumberg, M. S. (2002). *Bodyheat: Temperature and Life on Earth.* Cambridge, Massachusetts: Harvard University Press.

Blythman, J. (2004). *Shopped: the Shocking Truth about Supermarkets.* London: Fourth Estate.

Bodenhausen, G. V., Gabriel, S. & Lineberger, M. (2000). 'Sadness and Susceptibility to Judgmental Bias: The Case of Anchoring'. *Psychological Science* 11(4): 320–3.

Boland, J. E., Chua, H. F. & Nisbett, R. E. (2005). 'How We See it: Culturally Different Eye Movement Patterns over Visual Scenes'. In Rayner, K., Shen, D., Bai, X. & Yan, G. (eds.) *Cognitive and Cultural Influences on Eye Movements.* Tianjin, China: People's Press/Psychology Press. 363–78.

Bordwell, D. (2006). *The Way Hollywood Tells It. Story and Style in Modern Movies.* Berkeley: University of California Press.

Bornstein, B. H., Marcus, D. & Cassidy, W. (2000). 'Choosing a Doctor: an Exploratory Study of Factors Influencing Patients' Choice of a Primary Care Doctor'. *Journal of Evaluation in Clinical Practice* 6(3): 255–62.

Boston Society for Medical Improvement (1849). *Records of Meetings* (Vol.VI) Countway Library Mss, B MS b.92.2

Boxtel, G. J. M., Denissen, A., Jäger, M., Vernon, D., Dekker, M. K. J., Mihajlovic, V. & Sitskoorn, M. M. (2011). 'A Novel Self-Guided Approach to Alpha Activity Training'. *International Journal of Psychophysiology* (In Press).

Boy, F., Evans, C. J., Edden, R. A. E., Singh, K. D., Husain, M. & Sumner, P. (2010). 'Individual Differences in Subconscious Motor Control

Predicted by GABA Concentration in SMA'. *Current Biology* 20(19): 1779–85.

Boy, F., Husain, M. & Sumner, P. (2010). 'Unconscious Inhibition Separates Two Forms of Cognitive Control'. *PNAS* 107(24): 11134–9.

Brashers, D. E. (2001). 'Communication and Uncertainty Management'. *Journal of Communication* 51(3): 477–97.

Brass, M., Ruby, P. & Spengler, S. 'Inhibition of Imitative Behaviour and Social Cognition'. *Philosophical Transactions of the Royal Society B: Biological Sciences* 364(1528): 2359–67.

Brewer, M. B. & Kramer, R. M. (1985). 'The Psychology of Intergroup Attitudes and Behavior'. *Annual Review of Psychology* 36, 219–43.

Briers, B., Pandelaere, M., Dewitte, S. & Warlop, L. (2006). 'Hungry for Money: The Desire for Caloric Resources Increases the Desire for Financial Resources and Vice Versa'. *Psychological Science* 17(11): 939–43.

Broca, P. (1861). 'Perte de la Parole, Ramollissement Chronique et Destruction Partielle du Lobe Antérieur Gauche'. *Bulletin de la Société d'Anthropologie* 2: 235–38.

Broca, P. (1861). 'Nouvelle Observation d'Aphémie Produite par une Lésion de la Moitié Postérieure des Deuxième et Troisième Circonvolution Frontales Gauches'. *Bulletin de la Société Anatomique* 36: 398–407.

Brodie, B. B. & Shore, P. A. (1957). 'A Concept for a Role of Serotonin and Norepinephrine as Chemical Mediators in the Brain'. *Annals of New York Academy of Sciences* 66(3): 631–42.

Brüne, M., Schöbel, A., Karau, R., Faustmann, P. M., Dermietzel, R., Juckel, G. & Petrasch-Parwez, E. (2011). Neuroanatomical Correlates of Suicide in Psychosis: The Possible Role of von Economo Neurons. *PLoS One*, 6(6), e20936, 1–6.

Buckholtz, J. W., Treadway, M. T., Cowan, R. L., Woodward, N. D., Li, R., Sib Ansari, M., Baldwin, R. M., Schwartzman, A. N., Shelby, E. S., Smith, C. E., Kessler, R. M. & Zald, D. H. (2010). 'Dopaminergic Network Differences in Human Impulsivity'. *Science* 329(5991): 532.

Burns, J. M. & Swerdlow, R. H. (2003). 'Right Orbitofrontal Tumor with Pedophilia Symptom and Constructional Apraxia Sign'. *Archives of Neurology* 60(3), 437–40.

Burton Jr, V. S., Cullen, F. T., Evans, T. D., Alarid, L. F. & Dunaway, R. G. (1998). 'Gender, Self-Control and Crime'. *Journal of Research in Crime and Delinquency* 35(2): 123–47.

Bush, G. (2010). 'Attention-Deficit/Hyperactivity Disorder and Attention Networks'. *Neuropsychopharmacology* 35(1): 278300.

Buss, A. H. & Plomin, R. (1975). *A Temperament Theory of Personality Development*. New York: John Wiley & Sons.

Butnik, S. M. (2005). 'Neurofeedback in Adolescents and Adults with Attention Deficit Hyperactivity Disorder'. *Journal of Clinical Psychology*, 61(5): 621–5.

Camus, A. (1955). 'An Absurd Reasoning'. In O'Brien, J. (Ed. & Trans.), *The Myth of Sisyphus and Other Essays*. New York: Vintage Books 94.

Cardinal, R. N., Parkinson, J. A. & Everitt, B. J. (2002). 'Emotion and Motivation: The Role of the Amygdala, Ventral Striatum, and Prefrontal Cortex'. *Neuroscience and Biobehavioural Reviews* 26(3): 321–52.

Carew, T. J., Goldberg, M. E. & Marder, E. (2008). Brain Briefing: Mirror Neurons. *Society for Neuroscience*.

Carey, W. B., Fox, M. & McDevitt, S. C. (1977). 'Temperament as a Factor in Early School Adjustment'. *Pediatrics* 60(4 Pt 2): 621–4.

Carmon, Z. & Ariely, D. (2000). 'Focusing on the Forgone: How Value Can Appear So Different to Buyers and Sellers'. *Journal of Consumer Research* 27(3): 360–70.

Carré, J. M. & McCormick, C. M. (2008). 'In Your Face: Facial Metrics Predict Aggressive Behaviour in the Laboratory and in Varsity and Professional Hockey Players'. *Proceedings of the Royal Society B: Biological Sciences* 275(1651): 2651–6.

Carter, T. J., Ferguson, M. J. & Hassin, R. R. (2011). 'A Single Exposure to the American Flag Shifts Support Toward Republicanism up to 8 Months Later'. *Psychological Science* 22(8): 1011–18.

Carus, P. (1969). *The History of the Devil*. New York: Lands End Press.

Carver, C. S., Johnson, S. L., Joormann, J., Kim, Y. & Nam, J. Y. (2011). 'Serotonin Transporter Polymorphism Interacts with Childhood Adversity to Predict aspects of Impulsivity'. *Psychological Science* 22(5): 589–95.

Cattell, H. E. P. & Mead, A. D. (2008). 'The Sixteen Personality Factor

Questionnaire' (16PF). In Boyle, G., Matthews, G., Saklofske, D. H. (Eds.) *The SAGE Handbook of Personality Theory and Assessment; Vol 2 Personality Measurement and Testing.* Los Angeles, CA: Sage, 135–78.

Chamberlain, S. R. & Sahakian, B. J. (2007). 'The Neuropsychiatry of Impulsivity'. *Current Opinion in Psychiatry* 20(3): 255–61.

Chapman, G. B. & Johnson, E. J. (1994). 'The Limits of Anchoring'. *Journal of Behavioural Decision Making* 7(4): 22342.

Charach, A., Ickowicz, A. & Schachar, R. (2004). 'Stimulant Treatment Over Five Years: Adherence, Effectiveness, and Adverse Effects'. *Journal of American Academic Child Adolescent Psychiatry* 43(5): 55967.

Charach, A., Figueroa, M., Chen, S., Ickowicz, A. & Schachar, R. (2006). 'Stimulant Treatment Over 5 years: Effects on Growth'. *Journal of American Academic Child Adolescent Psychiatry* 45(4): 415–21.

Chester, J. & Montgomery, K. (2007). 'American University Interactive Food & Beverage Marketing: Targeting Children and Youth in the Digital Age'. Report from Berkeley Media Studies Group, May.

Chester, J. & Montgomery, K. C. (2008). 'Interactive Food and Beverage Marketing: Targeting Children and South in the Digital Age'. *Working Papers.*

Chester, J. & Montgomery, K. C. (2011). 'Digital Food Marketing to Children and Adolescents: Problematic Practices and Policy Interventions'. *Report for the National Policy & Legal Analysis Network to Prevent Childhood Obesity.*

Chester, J., Cheyne, A. & Dorfman, L. (2011). 'Peeking Behind the Curtain: Food and Marketing Industry Research Supporting Digital Media Marketing to Children and Adolescents'. *Center for Digital Democracy and Berkley Media Studies Group.*

Christoff, K., Gordon, A. & Smith, R. (2011). 'The Role of Spontaneous Thought in Human Cognition'. In Vartanian, O., Mandel. D. R. (Eds.) *Neuroscience of Decision Making.* Psychology Press.

Clark, R. D. & Hatfield, E. (1989). 'Gender Differences in Receptivity to Sexual Offers'. *Journal of Psychology Human Sexuality* 2(1): 39–45.

Cleeremans, A. (2005). 'Computational Correlates of Consciousness'. *Progress in Brain Research* 150: 81–98.

Cohen, M. X., Bour, L., Mantione, M., Figee, M., Vink, M., Tijssen, M. A., Rootselaar, A. F., Munckhof, P., Schuurman, P. & Denys, D. (2012).

'Top-Down Directed Synchrony from Medial Frontal Cortex to Nucleus Accumbens during Reward Anticipation'. *Human Brain Mapping* 33(1): 246–52.

Coll, A. P., Farooqi, I. S. & O'Rahilly, S. (2007). 'The Hormonal Control of Food Intake'. *Cell* 129(2): 251–62.

Coll, A. P., Yeo, G. S., Farooqi, I. S. & O'Rahilly, S. (2008). 'SnapShot: The Hormonal Control of Food Intake'. *Cell* 135(3): 572: e1–2.

Conley, T. D., Moors, A. C., Matsick, J. L., Ziegler, A. & Valentine, B. A. (2011). 'Women, Men, and the Bedroom: Methodological and Conceptual Insights That Narrow, Reframe, and Eliminate Gender Differences in Sexuality'. *Current Directions in Psychological Science* 20(5): 296–300.

Cooley, C. H. (1998). *On Self and Social Organization* (Hans-Joachim, S., Ed.). Chicago: University of Chicago 20–2.

Cooling, J. & Blundell, J. E. (2001). 'High-Fat and Low-Fat Phenotypes: Habitual Eating of High- and Low-Fat Foods not Related to Taste Preference for Fat'. *European Journal of Clinical Nutrition* 55(11): 1016–21.

Corr, P. J., Pickering, A. D. & Gray, J. A. (1995). 'Personality and Reinforcement in Associative and Instrumental Learning'. *Personality and Individual Differences* 19(1): 47–71.

Corwin R. L., Grigson, P. S. & Grigson, P. S. (2002). 'Like Drugs for Chocolate: Separate Rewards Modulated by Common Mechanisms?' *Physiology & Behaviour* 76(3): 389–95.

Cota, D., Tschop, M. H., Horvath, T. L. & Levine, A. S. (2006). 'Cannabinoids, Opioids and Eating Behavior: The Molecular Face of Hedonism?' *Brain Research Reviews* 51(1): 85–107.

Coward, R. (1984). *Female Desire.* London: Paladin.

Csatho, A., Osvath, A., Bicsak, E., Karadi, K., Manning, J. & Kallai, J. (2003). 'Sex Role Identity Related to the Ratio of Second to Fourth Digit Length in Women'. *Biological Psychology* 62(2): 147–56.

Csibra, G. (2005). Mirror Neurons and Action Observation: Is Simulation Involved?

Dallas, E. S. (1886). *The Gay Science,* London: Chapman & Hall.

Damasio, H., Grabowski, T., Frank, R., Galaburda, A. M. & Damasio, A. R. (1994). 'The Return of Phineas Gage: Clues about the Brain from the Skull of a Famous Patient'. *Science* 264(5162): 1102–5.

Damasio, A. R. (1994). *Descartes' Error: Emotion, Reason and the Human Brain*. New York: G. P. Putman.

Damasio, A. R., (1995) *Descartes' Error: Emotion, Reason and the Human Brain*. New York: Avon Books.

Daruna, J. H. & Barnes, P. A. (1993). 'A Neurodevelopmental View of Impulsivity'. In McCown, W. G., Johnson, J. L., Shure, M. B. (Eds.) *The Impulsive Client: Theory, Research and Treatment*. Washington, DC: American Psychological Association.

Davis, C. & Fox, J. (2008). 'Sensitivity to Reward and Body Mass Index (BMI): Evidence for a Non-Linear Relationship'. *Appetite* 50(1): 43–9.

Dawkins, R. (2011). 'The Tyranny of the Discontinuous Mind'. *New Statesman*, 19th Dec – 1st Jan: 54–7.

De Gennaro, R. (1979). *Research Libraries Enter the Information Age*. Seventh R. R. Bowker Memorial Lecture, New York: 13th November.

DellaSanta-Percy, C. 'Interdisciplinary Fitness Interview – IFI and IFI-R'. *Unpublished Paper*.

DelParigi, A., Chen, K., Salbe, A. D., Reiman, E. M., & Tataranni, P. A. (2005). 'Sensory Experience of Food and Obesity: A Positron Emission Tomography Study of the Brain Regions Affected by Tasting a Liquid Meal After a Prolonged Fast'. *Neuroimage* 24(2): 436–43.

DelParigi, A., Pannacciulli, N., Le, D. N. & Tataranni, P. A. (2005). 'In Pursuit of Neural Risk Factors for Weight Gain in Humans'. *Neurobiology of Aging, 26 Supplement (1)* 50–5.

Demont, L. (1933). *A Concise Dictionary of Psychiatry and Medical Psychology*. Philadelphia: Lippincott.

Dennett, D. C. (1992). *Consciousness Explained*. Back Bay Books.

Depue, R. A., Collins, P. F. (1999). 'Neurobiology of the Structure of Personality: Dopamine, Facilitation of Incentive Motivation, and Extraversion'. *Behavioral and Brain Sciences* 22(3): 491–569.

Derbyshire, D. (2004) 'They Have Ways of Making You Spend'. *The Telegraph* December (Sourced Online January 2012).

De Renzi, E., Cavalleri, F. & Facchini, S. (1996). 'Imitation and Utilisation Behaviour'. *Journal of Neurology, Neurosurgery and Psychiatry* 61: 396–400.

Desmond, A. & Moore, E. (1991). *Darwin* London: Michael Joseph. 257.

Dewall, C., Baumeister, R. F., Stillman, T. & Gailliot, M. 'Violence

Restrained: Effects of Self-Regulation and its Depletion on Aggression'. *Journal of Experimental Social Psychology* 43(1): 62–76.

Dichter, E. (1960). *The Strategy of Desire* New York: Boardman.

Dickman, S. J. (1990). 'Functional and Dysfunctional Impulsivity: Personality and Cognitive Correlates'. *Journal of Personality and Social Psychology* 58(1): 95–102.

Diener, E., Emmons, R. A., Larsen, R. J. & Griffin, S. (1985). 'The Satisfaction with Life Scale'. *Journal of Personality Assessment* 49(1): 71–5.

Dietrich, M. O. & Horvath, T. L. (2009). 'Feeding Signals and Brain Circuitry'. *European Journal of Neuroscience* 30(9): 1688–96.

Dijksterhuis, A. (2004). 'Think Different: The Merits of Unconscious Thought in Preference Development and Decision Making'. *Journal of Personality and Social Psychology* 87(5): 586–98.

Dijksterhuis, A., Smith, P. K., Van Baaren, R. B. & Wigboldus, D. H. J. (2005). 'The Unconscious Consumer: Effects of Environment on Consumer Behavior'. *Journal of Consumer Psychology* 15(3): 193–202.

Dijksterhuis, A. & Van Olden, Z. (2006). 'On the Benefits of Thinking Unconsciously: Unconscious Thought can Increase Post-Choice Satisfaction'. *Journal of Experimental Social Psychology* 42(5): 627–31.

Dijksterhuis, A. & Nordgren, L. F. (2006). 'A Theory of Unconscious Thought'. *Perspectives on Psychological Science* 1(2): 95–109.

Dörner, D. (1996). *The Logic of Failure.* New York: Metropolitan Books. 28–34.

Dreber, A., Apicella, C. L., Eisenberg, D. T. A., Garcia, J. R., Zamore, R. S., Lum, J. K. & Campbell, B. (2009). 'The 7R Polymorphism in the Dopamine Receptor D4 Gene (DRD4) is Associated with Financial Risk Taking in Men'. *Evolution and Human Behavior* 30(2): 85–92.

Dreber, A., Rand, D. G., Garcia, J. R., Wernerfelt, N., Koji Lum, J. & Zeckhauser, R. (2010). 'Dopamine and Risk Preferences in Different Domains'. *Harvard Kennedy School, Faculty Research Working Paper Series.*

Drury, J., Reicher, S. (2000). 'Collective Action and Psychological Change: The Emergence of New Social Identities'. *British Journal of Social Psychology* 39(4) 579–604.

Duckworth, A. L. & Seligman, M. E. (2005). 'Self-Discipline Outdoes IQ in Predicting Academic Performance of Adolescents'. *Psychological Science* 16(12): 939–44.

Duclos, A., Peix, J-L., Colin, C., Kraimps, J-L., Menegaux, F., Pattou, F., Sebag, F., Touzet, S., Bourdy, S., Voirin, N. & Lifante, J–C. (2012). 'Influence of Experience on Performance of Individual Surgeons in Thyroid Surgery: Prospective Cross Sectional Multicentre Study'. *British Medical Journal* 344.

Ecomist, Findings of Recent Research of Fragrancing in Retail Environments. *(www.ecomistsystems.com)*

Elias, M. (2005). 'So Much Media, So Little Attention Span'. *USA Today*, 3rd November.

Ellis, B. J. (1992). 'The Evolution of Sexual Attraction: Evaluative Mechanisms in Women'. In Barkow, J., Cosmides, L., Tooby, J., (Eds.). *The Adapted Mind*, New York: Oxford University Press 267–88.

Ellis, L. K., Rothbart, M. K., Posner, M. I. (2004). 'Individual Differences in Executive Attention Predict Self-Regulation and Adolescent Psychosocial Behaviours'. *Annals of the New York Academy of Sciences*, 1021, 337–40.

Englich, B. & Mussweiler, T. (2001). 'Sentencing under Uncertainty: Anchoring Effects in the Courtroom'. *Journal of Applied Social Psychology* 31(7): 1535–51.

English, H. (1928). *A Student's Dictionary of Psychological Terms.* Yellow Springs OH: Antioch Press.

Erlanson-Albertsson, C. (2005). 'Sugar Triggers Our Reward-System. Sweets Release Opiates Which Stimulates the Appetite for Sucrose – Insulin Can Depress It'. *Lakartidningen* 192(21): 1620–2, 1625, 1627.

Erlanson-Albertsson, C. (2005). 'How Palatable Food Disrupts Appetite Regulation'. *Basic Clinical Pharmacology and Toxicology,* 97(2): 61–73.

Eskine, K. J., Kacinik, N. A. & Prinz, J. J. (2011). 'A Bad Taste in the Mouth: Gustatory Disgust Influences Moral Judgment'. *Psychological Science* 22(3): 295–9.

Eslinger, P. J. & Damasio, A. R. (1985). 'Severe Disturbance of Higher Cognition after Bilateral Frontal Lobe Ablation: Patient EVR'. *Neurology* 35(12): 1731–41.

Eslinger, P. J., Dennis, K., Moore, P., Antani, S., Hauck, R. & Grossman,

M. (2005). 'Metacognitive Deficits in Frontotemporal Dementia'. *Journal of Neurology, Neurosurgery and Psychiatry* 76(12): 1630–5.

Evans, J. St. B. T. (2003). 'In Two Minds: Dual-Process Accounts of Reasoning'. *Trends in Cognitive Sciences* 7(10): 454–9.

Evans, J. St. B. T. (2008). 'Dual-Processing Accounts of Reasoning, Judgment, and Social Cognition'. *Annual Review of Psychology* 59: 255–78.

Evans, J. St. B. T., Barston, J. L., Pollard, P. (1983). 'On the Conflict Between Logic and Belief in Syllogistic Reasoning'. *Memory and Cognition* 11(3): 295–306.

Evans, J. St. B. T., Over, D. E. (1996). *Rationality and Reasoning*. Psychology Press.

Evenden, J. (1999). 'Impulsivity: A Discussion of Clinical and Experimental Findings'. *Journal of Psychopharmacology* 13(2): 180–92.

Executive Leadership Foundation. The Sigmoid Curve and the Paradox of Success.

Eysenck, H. J. & Eysenck, M. W. (1985). *Personality and Individual Differences: a Natural Science Approach*. New York: Plenum Press.

Faraone, S. V., Biederman, J. & Mick, E. (2006). 'The Age-Dependent Decline of Attention Deficit Hyperactivity Disorder: A Meta-Analysis of Follow-Up Studies'. *Psychological Medicine* 36(2): 159–65.

Farrington, D. P., Bowen, S., Buckle, A., Burns-Howell, T., Burrows, J. & Speed, M. (1993). 'An Experiment on the Prevention of Shoplifting'. *Crime Prevention Studies*, 1, 92–119.

Fazel, S. & Danesh, J. (2002). 'Serious Mental Disorder in 23000 Prisoners: A Systematic Review of 62 Surveys'. *Lancet* 359(9306): 545–50.

Feldman, R., Greenbaum, C. W. & Yirmiya, N. (1999). 'Mother-Infant Affect Synchrony as an Antecedent of the Emergence of Self-Control'. *Developmental Psychology* 35(1): 223–31.

Ferrier, D. (1878). 'The Goulstonian Lectures on the Localisation of Cerebral Disease'. *British Medical Journal* i 397–442, 443–7.

Feynman, R. (1988). *What Do You Care What Other People Think?* New York: W. W. Norton & Co. Inc. 138.

Figner, B., Mackinlay, R. J., Wilkening, F. & Weber, E. U. (2009). 'Affective and Deliberative Processes in Risky Choice: Age Differences in Risk

Taking in the Columbia Card Task'. *Journal of Experimental Psychology: Learning, Memory, and Cognition* 35(3): 709–30.

Figner, B. & Weber, E. U. (2011). 'Who Takes Risks When and Why? Determinants of Risk Taking'. *Current Directions in Psychological Science* 20(4): 211–16.

Fingelkurts, A. A., Fingelkurts, A. A., Bagnato, S., Boccagni, C. & Galardi, G. (2011). 'Toward Operational Architectonics of Consciousness: Basic Evidence from Patients with Severe Cerebral Injuries'. *Cognitive Processing,* (Online Publication).

Fink, B., Grammer, K. & Matts, P. (2006). 'Visible Skin Color Distribution Plays a Role in the Perception of Age, Attractiveness, and Health in Female Faces'. *Evolution and Human Behavior* 27(6): 433–42.

Finkel, E. J. & Eastwick, P. W. (2009). 'Arbitrary Social Norms Influence Sex Differences in Romantic Selectivity'. *Psychological Science* 20(10): 1290–5.

Fischhoff, B. (2007). 'An Early History of Hindsight Research'. *Social Cognition* 25(1): 10–13.

Fischhoff, B. & Beyth, R. (1975). 'I Knew it Would Happen: Remembered Probabilities of Once-Future Things'. *Organizational Behaviour and Human Performance,* 13(1), 1–16.

Fischhoff, B. (1982). 'For Those Condemned to Study the Past: Heuristics and Biases in Hindsight'. In Kahneman, D., Slovic, P., Tversky, A., (Eds.) *Judgement under Uncertainty: Heuristics and Biases.* Cambridge: Cambridge University Press 332 – 51.

Fisher, D. J. (1988). *Rules of Thumb.* Switzerland: Trans Tech Publications.

Fisher, T. D., Moore, Z. T. & Pittenger, M. J. (2011). 'Sex on the Brain? An Examination of Frequency of Sexual Cognitions as a Function of Gender, Erotophilia, and Social Desirability'. *Journal of Sex Research Advanced* (Online Publication).

Fiske, S. T., Cuddy, A. J. C. & Glick, P. (2007). 'Universal Dimensions of Social Cognition: Warmth and Competence'. *Trends in Cognitive Sciences,* 11(2): 77–83.

Fitzsimons, G. J., Hutchinson, J. W., Williams, P., Alba, J. W., Chartrand, T. L., Huber, J., Kardes, F. R., Menon, G., Raghubir, P., Russo, J. D., Shiv, B. & Tavaaoli, N. Y. (2002). 'Non-Conscious Influences on Consumer Choice'. *Marketing Letters* 13(3): 269–79.

Fitzsimons, G. M. & Bargh, J. A. (2003). 'Thinking of You: Non-Conscious Pursuit of Interpersonal Goals Associated With Relationship Partners'. *Journal of Personality and Social Psychology* 84(1): 148–64.

Fitzsimons, G. M., Chartrand, T. L. & Fitzsimons, G. J. (2008). 'Automatic Effects of Brand Exposure on Motivated Behavior: How Apple Makes You "Think Different"'. *Journal of Consumer Research* 35(1): 21–35.

Fletcher, W. (1992). *A Glittering Haze*. Henley-on-Thames: NTC Publications.

Fodor, J. (2001). *The Mind Doesn't Work That Way*. Cambridge, MA: MIT Press.

Fontaine, J. R. J., Scherer, K. R., Roesch, E. B. & Ellsworth, P. C. (2007). 'The World of Emotions is not Two-Dimensional'. *Psychological Science* 18(12): 1050–7.

Ford, C.S., Beach, F. A. (1951). *Patterns of Sexual Behavior*. New York: Harper & Row.

Fox, N. A. & Calkins, S. D. (2003). 'The Development of Self-Control of Emotion: Intrinsic and Extrinsic Influences'. *Motivation and Emotion* 27(1): 7–26.

Francis, C. (2002). 'Fast Food Nation: The Dark Side of the All-American Meal'. *Crop Science* 42(2): 657–a–8.

Friedman, M. (1976). *Price Theory*. Chicago: Aldine.

Fujita, K., Trope, Y., Liberman, N. & Levin-Sagi, M. (2006). 'Construal Levels and Self-Control'. *Journal of Personality and Social Psychology* 90(3): 351–67.

Fuller, G. N. & Burger, P. C. (1990). 'Nervus Terminalis (Cranial Nerve Zero) in the Adult Human'. *Clinical Neuropathology* 9(6): 279–83.

Furnham, A. & Baguma, P. (1994). 'Cross-Cultural Differences in the Evaluation of Male and Female Body Shapes'. *International Journal of Eating Disorders* 15(1): 81–9.

Furnham A., Lavancy, M. & McClelland, A. (2001). 'Waist to Hip Ratio and Facial Attractiveness: A Pilot Study'. *Personality and Individual Differences* 30(3): 491–502.

Furnham, A., McClelland, A. & Omer, L. (2003). 'A Cross-Cultural Comparison of Ratings of Perceived Fecundity and Sexual Attractiveness as a Function of Body Weight and Waist-to-Hip Ratio'. *Psychology, Health and Medicine* 8(2): 219–30.

Furnham, A., Swami, V. & Krupa, S. (2006). 'Body Weight, Waist-to-Hip Ratio and Breast Size Correlates of Ratings of Attractiveness and Health'. *Personality and Individual Differences* 41(3): 443–54.

Furnham, A. & Swami, V. (2007). 'Perception of Female Buttocks and Breast Size in Profile'. *Social Behavior and Personality* 35(1): 1–8.

Gailliot, M. T. & Baumeister, R. F. (2007). 'Self-Regulation and Sexual Restraint: Dispositionally and Temporarily Poor Self-Regulatory Abilities Contribute to Failures at Restraining Sexual Behavior'. *Personality and Social Psychology Bulletin* 33(2): 173–86.

Gailliot, M. T., Baumeister, R. F., DeWall, C. N., Maner, J. K., Plant, E. A., Tice, D. M., Brewer, L. E. & Schmeichel, B. J. (2007). 'Self-Control Relies on Glucose as a Limited Energy Source: Willpower is More Than a Metaphor'. *Journal of Personality and Social Psychology* 92(2): 325–36.

Galinsky, A. D. & Mussweiler, T. (2001). 'First Offers as Anchors: The Role of Perspective-Taking and Negotiator Focus'. *Journal of Personality and Social Psychology* 81(4): 657–69.

Gallese, V. (2001). 'The Shared Manifold Hypothesis: From Mirror Neurons to Empathy'. *Journal of Consciousness Studies* 8(5–7): 33–50.

Gallese, V. & Goldman, A. (1998). 'Mirror Neurons and the Simulation Theory of Mind-Reading'. *Trends in Cognitive Sciences* 2(12): 493–501.

Gallese, V., Eagle, M. N. & Migone, P. (2007). 'Intentional Attunement: Mirror Neurons and the Neural Underpinnings of Interpersonal Relations'. *Journal of the American Psychoanalytic Association* 55(1): 131–75.

Gardner, M. & Steinberg, L. (2005). 'Peer Influence on Risk Taking, Risk Preference, and Risky Decision Making in Adolescence and Adulthood: an Experimental Study'. *Developmental Psychology* 41(4): 625–35.

Giedd, J. N. (2007). 'The Teen Brain: Insights from Neuroimaging'. *Journal of Adolescent Health* 42(4): 335–43.

Gigerenzer, G. (2008). 'Why Heuristics Work'. *Perspectives on Psychological Science* 3(1): 20–9.

Gigerenzer, G. & Todd, P. M. (1999). *Simple Heuristics That Make Us Smart.* Oxford: Oxford University Press.

Gilden, D. L. & Marusich, L. R. (2009). 'Contraction of Time in Attention-Deficit Hyperactivity Disorder'. *Neuropsychology* 23(2): 265–9.

Gilovich, T., Medvec, V. H. & Medvec, V. H. (2000). 'The Spotlight Effect in Social Judgment: An Egocentric Bias in Estimates of the Salience of One's Own Actions and Appearance'. *Journal of Personality and Social Psychology* 78(2), 211–22.

Gino, M., Moore, D. A. & Bazerman, M. H. (2009). 'See No Evil: When We Overlook Other People's Unethical Behavior'. In Kramer, R. M., Tenbrunsel, A. E., Bazerman, M. H., (Eds.) *Social Decision Making: Social Dilemmas, Social Values, and Ethical Judgments* Psychology Press. 241–63.

Gino, F., Sharek, F. & Moore, D. A. (2011). 'Keeping the Illusion of Control under Control: Ceilings, Floors, and Imperfect Calibration'. *Organizational Behavior and Human Decision Processes* 114(2): 104–14.

Gleick, J. (1987). *Chaos: Making a New Science.* New York: Viking Penguin.

Godoy, R., Reyes-García, V., Huanca, T., Leonard, W. R., McDade, T., Tanner, S. & Seyfried, C. (2007). 'On the Measure of Income and the Economic Unimportance of Social Capital, Evidence from a Native Amazonian Society of Farmers and Foragers'. *Journal of Anthropological Research* 63(2): 239–60.

Goel, V. (2003). 'Evidence for Dual Neural Pathways for Syllogistic Reasoning'. *Psychologica* 32: 301–09.

Goel, V., Buchel, C., Frith, C. & Dolan, R. J. (2000). 'Dissociation of Mechanisms Underlying Syllogistic Reasoning'. *Neuroimage* 12(5): 504–14.

Golding, S. L., Roesch, R. & Schreiber, J. (1984). 'Assessment and Conceptualization of Competency to Stand Trial: Preliminary Data on the Interdisciplinary Fitness Interview'. *Law and Human Behaviour* 8(3/4): 321–34.

Goldstein, R. Z. & Volkow, N. D. (2011). 'Dysfunction of the Prefrontal Cortex in Addiction: Neuroimaging Findings and Clinical Implications'. *Nature Reviews Neuroscience* 12(11): 652–69.

Goslin, J., Dixon, T., Fischer, M. H., Cangelosi, A. & Ellis, R. (2012). 'Electrophysiological Examination of Embodiment in Vision and Action'. *Psychological Science* 23(2): 12–157.

Gosselin, N., Peretz, I., Johnsen, E. & Adolphs, R. (2007). 'Amygdala Damage Impairs Emotion Recognition from Music'. *Neuropsychologia* 45(2): 236–244.

Gray, J. A. (1973). 'Causal Theories of Personality and How to Test Them'. In Royce, J. R. (Ed.) *Multivariate Analysis and Psychological Theory.* London: Academic Press, 409–63

Greenberg, J. & Hollander, E. (2003). 'Brain Function and Impulsive Disorders'. *Psychiatric Times* (1st March).

Gregory, R. (2004). 'The Blind Leading the Sighted'. *Nature* 430(7002): 836.

Guerber, H. A. (1927). *Myths of Greece and Rome.* London: George G. Harrap & Co 313.

Gul, F. & Pesendorfer, W. (2001). 'Temptation and Self-Control'. *Econometrica* 69(6): 1403–35.

Haas L. F. (2001). 'Phineas Gage and the Science of Brain Localisation'. *Journal of Neurology, Neurosurgery and Psychiatry* 71(6): 761.

Han, Y. K., Morgan, G. A., Kotsiopulos, A. & Kang-Park, J. (1991). 'Impulse Buying Behavior of Apparel Purchasers'. *Clothing and Textiles Research Journal,* 9(3): 15–21.

Hanselmann, M. & Tanner, C. (2008). 'Taboos and Conflicts in Decision Making: Sacred Values, Decision Difficulty, and Emotions'. *Judgment and Decision Making* 3(1): 51–63.

Hanslmayr, S., Gross, J., Klimesch, W. & Shapiro, K. L. (2011). 'The Role of Alpha Oscillations in Temporal Attention'. *Brain Research Reviews* 67(1–2): 331–43.

Hardin. J. (2009). *Kids: 'They're not Going to Outlive Their Parents'.* News & Record, Greensboro, North Carolina.

Hardy, G. H. (1940). *Ramanujan.* Cambridge: Cambridge University Press.

Harinck, F., Van Dijk, E., Van Beest, I. & Mersmann, P. (2007). 'When Gains Loom Larger than Losses: Reversed Loss Aversion for Small Amounts of Money'. *Psychological Science* 18(12): 1099–105.

Harlow, J. M. (1848). 'Passage of an Iron Rod Through the Head'. *Boston Medical and Surgical Journal,* 39, 389–93. [Reprinted in: (1999) *The Journal of Neuropsychiatry and Clinical Neurosciences* 11(2): 281–3]

Harlow, J. M. (1849). 'Letter in Medical Miscellany'. *Boston Medical and Surgical Journal,* 39, 506–7. [Copy in Macmillan, M. (2000). *An Odd Kind of Fame. Stories of Phineas Gage.* Cambridge, Mass: The MIT Press].

Harlow, J. M. (1868). 'Recovery from the Passage of an Iron Bar through the Head'. *Publications of the Massachusetts Medical Society*, 2, 327–347. [reprinted in: (1993) *History of Psychiatry* 4(14): 274–81].

Harmon-Jones, E. & Gable, P. A. (2009). 'Neural Activity Underlying the Effect of Approach-Motivated Positive Affect on Narrowed Attention'. *Psychological Science* 20(4): 406–9.

Harris, K. D. & Thiele, A. (2011). 'Cortical State and Attention'. *Nature Reviews Neuroscience*, 12(9): 509–23.

Haslam, S. A. & Reicher, S. (2007). 'Beyond the Banality of Evil: Three Dynamics of an Interactionist Social Psychology of Tyranny'. *Personality and Social Psychology Bulletin* 33(5): 615–22.

Hassin, P. R., Bargh, J. A. & Zimerman, S. (2009). 'Automatic and Flexible: The Case of Non-Conscious Goal Pursuit'. *Social Cognition* 27(1): 20–36.

Hatemi, P. K., McDermott, R., Eaves, L. J., Kendler, K. S. & Neale, M. C. *Fear Dispositions and their Relationship to Political Preferences*. Unpublished Mss, Pennsylvania State University.

Hausman, A. (2000). 'A Multi-Method Investigation of Consumer Motivations in Impulse Buying Behavior'. *Journal of Consumer Marketing* 17(5): 403–17.

Hell, W., Gigerenzer, G., Gauggel, S., Mall, M. & Müller, M. (1988). 'Hindsight Bias: An Interaction of Automatic and Motivational Factors?' *Memory & Cognition* 16(6): 533–8.

Herbert, W. (2010). 'On Second Thought'. The Denver Post, 28th October.

Herz, R. (2007). *The Scent of Desire: Discovering Our Enigmatic Sense of Smell*. William Morrow & Co.

Heyes, C. (2010). 'Where Do Mirror Neurons Come From?' *Neuroscience and Biobehavioral Reviews* 34(4): 575–83.

Hinsley, F. H. & Stripp, A. (1993). *Codebreakers: The Inside Story of Bletchley Park* Oxford: Oxford University Press.

Hinslie, L. & Shatzky, J. (1940). *Psychiatric Dictionary*. New York: Oxford University Press.

Hippocrates. (Adams, F. 1886 Trans.). *The Genuine Works of Hippocrates*, 2, 344–5.

Hirt, E. (1902). *Die Temperamente* (The Temperaments). Leipzig: Barth.

Hoffman, Y., Bein, O. & Maril, A. (2011). 'Explicit Memory for Unattended Words: The Importance of Being in the "No"'. *Psychological Science* 22(12): 1490–3.

Hofmann, W., Rauch, W. & Gawronski, B. (2007). 'And Deplete Us Not into Temptation: Automatic Attitudes, Dietary Restraint, and Self-Regulatory Resources as Determinants of Eating Behavior'. *Journal of Experimental Social Psychology* 43(3): 497–504.

Hofmann, W., Strack, F. & Deutsch, R. (2008). 'Free to Buy? Explaining Self-Control and Impulse in Consumer Behavior'. *Journal of Consumer Psychology* 18(1): 22–6.

Hofmann, W., Friese, M. & Strack, F. (2009). 'Impulse and Self-Control from a Dual-Systems Perspective'. *Perspectives on Psychological Science* 4(2): 162.

Hofmann, W., Baumeister, R. F., Förster, G. & Vohs, K. D. (2011). 'Everyday Temptations: An Experience Sampling Study of Desire, Conflict, and Self-Control'. *Journal of Personality and Social Psychology* (Online Publication).

Holland, P. (1997). *The Television Handbook*. London: Routledge.

Holland, R. W., Hendriks, M. & Aarts, H. (2005). 'Smells like Clean Spirit: Non-Conscious Effects of Scent on Cognition and Behavior'. *Psychological Science* 16(9): 689–93.

Hume, L., Dodd, C. A. & Grigg, N. P. (2003). 'In-Store Selection of Wine – No Evidence for the Mediation of Music?' *Perceptual and Motor Skills* 96(3 Pt 2): 1252–4.

Humphery, K. (1998). *Shelf Life: Supermarkets and the Changing Cultures of Consumption*. Cambridge: Cambridge University Press.

Hsu, L. M., Chung, J. & Langer, E. J. (2010). 'The Influence of Age-Related Cues on Health and Longevity'. *Perspectives on Psychological Science*, 5(6): 632–48.

IJzerman, H. I. & Semin, G. R. (2009). 'The Thermometer of Social Relations'. *Psychological Science* 20(10): 1214–20.

Inzlicht, M., McKay, L. & Aronson, J. (2006). 'Stigma as Ego-Depletion: How Being the Target of Prejudice Affects Self-Control'. *Psychological Science* 17(3): 262–9.

Inzlicht, M. & Gutsell, J. N. (2007). 'Running on Empty: Neural Signals for Self-Control Failure'. *Psychological Science* 18(11): 933–7.

Jacobson, L., Ezra, A., Berger, U. & Lavidor, M. (2011). 'Modulating Oscillatory Brain Activity Correlates of Behavioral Inhibition Using Transcranial Direct Current Stimulation'. *Clinical Neurophysiology* (In Press).

Jacoby, L. L., Woloshyn, V. & Kelley, C. (1989). 'Becoming Famous Without Being Recognized: Unconscious Influences of Memory Produced by Dividing Attention'. *Journal of Experimental Psychology: General* 118(2): 115–25.

Jackson, J. B. S. (1849). 'Medical Cases' (Vol.4, Cases Number 1358–1929, 720, 610). Countway Library Mss, H MS b 72.4. [Copy of relevant sections in Macmillan, M., (2000). *An Odd Kind of Fame. Stories of Phineas Gage.* Cambridge, Mass: The MIT Press].

Jackson, J. B. S. (1870). *A Descriptive Catalogue of the Warren Anatomical Museum.* Boston. MA: Williams.

James, G. & Koehler, D. J. (2011). 'Banking on a Bad Bet: Probability Matching in Risky Choice Is Linked to Expectation Generation'. *Psychological Science* 22(6): 707–11.

James, W. (1890). *The Principles of Psychology.* Cambridge, Mass: Harvard University Press. 119–20.

James, W. (1890). *The Principles of Psychology.* New York: H. Holt and company. 291–2.

Janis, I. L. (1982). 'Decision Making Under Stress'. In Goldberger, L., Breznitz, S. (Eds.) *Handbook of Stress: Theoretical and Clinical Aspects.* New York: The Free Press. 73.

Janis, I. L. & Mann, L. (1977). *Decision Making: A Psychological Analysis of Conflict, Choice, and Commitment.* New York: Free Press.

Jasieńska, G., Ziomkiewicz, A., Ellison, P. T., Lipson, S. F. & Thune, I. (2004). 'Large Breasts and Narrow Waists Indicate High Reproductive Potential in Women'. *Proceedings of the Royal Society B: Biological Sciences* 271(1545): 1213–17.

Jaśkowski, P., Skalska, B. & Verleger, R. (2003). 'How the Self Controls Its "Automatic Pilot" when Processing Subliminal Information'. *Journal of Cognitive Neuroscience* 15(6): 911–20.

Job, V., Dweck, C. S. & Walton, G. M. (2010). 'Ego Depletion – Is It All in Your Head? Implicit Theories About Willpower Affect Self-Regulation'. *Psychological Science* 21(11): 1686–93.

John, O. P. Naumann, L. P. (2010). 'Surviving Two Critiques by Block? The Resilient Big Five Have Emerged as the Paradigm for Personality Trait Psychology'. *Psychological Inquiry* 21(1): 44–9.

Johnson, P. M. & Kenny, P. J. (2010). 'Dopamine D2 Receptors in Addiction-Like Reward Dysfunction and Compulsive Eating in Obese Rats'. *Nature Neuroscience* 13(5): 635–41.

Johnstone, S. J., Roodenrys, S., Blackman, R., Johnston, E., Loveday, K., Mantz, S. & Barratt, M. F. (2011). 'Neurocognitive Training for Children with and without AD/HD'. *Attention Deficit and Hyperactivity Disorders*, (Online Publication).

Jones, B. C., Little, A. C., Burt, D. M. & Perrett, D. I. (2004). 'When Facial Attractiveness is Only Skin Deep'. *Perception* 33(5): 569–76.

Joyce, J. (1992). *Ulysses*. Harmondsworth: Penguin Books.

Kacen, J. J. & Lee, J. A. (2002). 'The Influence of Culture on Consumer Buying Behaviour'. *Journal of Consumer Psychology* 12(2): 163–78.

Kahneman, D. (2011). *Thinking, Fast and Slow*. New York: Farrar, Straus and Giroux.

Kahneman, D. & Tversky, A. (1979). 'Prospect Theory: An Analysis of Decisions Under Risk'. *Econometrica* 47(2): 263–91.

Kahneman, D. & Tversky, A. (1984). 'Choices, Values, and Frames'. *American Psychologist* 39(4): 341–50.

Kahneman, D., Knetsch, J. K. & Thaler, R. H. (1991). 'Anomalies: The Endowment Effect, Loss Aversion, and Status Quo Bias'. *The Journal of Economic Perspectives* 5(1): 193–206.

Kahneman, D., Fredrickson, B. L., Schreiber, C. A. & Redelmeier, D. A. (1993). 'When More Pain Is Preferred to Less: Adding a Better End'. *Psychological Science* 4(6): 401–5.

Kahneman, D. & Frederick, S. (2002). 'Representativeness Revisited: Attribute Substitution in Intuitive Judgement'. In *Heuristics and Biases: The Psychology of Intuitive Judgment*, In Gilovich, T., Griffin, D., Kahneman, D. (Eds.). Cambridge, UK: Cambridge University Press, 49–81.

Kamin, K. & Rachlinski, J. (1995). 'Ex Post ≠ Ex Ante: Determining Liability in Hindsight'. *Law and Human Behavior* 19(1): 89–104.

Kandinsky, V. (1885). *Kritische und klinische Betrachtungen im Gebiete der Sinnestäuschungen. Erste und zweite Studie*. Berlin: Verlag von Friedländer und Sohn.

Kassam, K. S., Koslov, K. & Mendes, W. B. (2009). 'Decisions Under Distress: Stress Profiles Influence Anchoring and Adjustment'. *Psychological Science,* 20(11): 1394–9.

Kelley, A. E., Bakshi, V. P., Haber, S. N., Steininger, T. L., Will, M. J. & Zhang, M. (2002). 'Opioid Modulation of Taste Hedonics Within the Ventral Striatum'. *Physiology & Behaviour* 76(3): 365–77.

Kelley, A. E., Schiltz, C. A. & Landry, C. F. (2005). 'Neural Systems Recruited by Drug- and Food-Related Cues: Studies of Gene Activation in Corticolimbic Regions'. *Physiology & Behavior* 86(1–2): 11–14.

Kelly, D. (2011) *Yuck! The Nature and Moral Significance of Disgust.* Cambridge, Mass: The MIT Press. 65.

Keysers, C., Kohler, E., Umiltà, M. A., Nanetti, L., Fogassi, L. & Gallese, V. (2003). Audiovisual Mirror Neurons and Action Recognition. *Experimental Brain Research* 154(4): 628–36.

Kirsch, P., Esslinger, C., Chen, Q., Mier, D., Lis, S., Siddhanti, S., Gruppe, H., Mattay, V. S., Gallhofer, B. & Meyer-Lindenberg, A. (2005). 'Oxytocin Modulates Neural Circuitry for Social Cognition and Fear in Humans'. *Journal of Neuroscience* 25(49): 11489–93.

Kivetz, R. & Keinan, A. (2006). 'Repenting Hyperopia: An Analysis of Self-Control Regrets'. *Journal of Consumer Research* 33(3), 273–82.

Klein, O., Spears, R. & Reicher, S. (2007). 'Social Identity Performance: Extending the Strategic Side of SIDE'. *Personality and Social Psychology Review* 11(1): 28:45.

Knetsch, J. L. (1989). 'The Endowment Effect and Evidence of Non-Reversible Indifference Curves'. *American Economic Review* 79(5): 1277–84.

Koehler, J. J. (2001). 'The Psychology of Numbers in the Courtroom: How to Make DNA-Match Statistics Seem Impressive or Insufficient'. *Southern California Law Review,* 74(5) 1275–306.

Koehler, J. J. (2011). 'If the Shoe Fits They Might Acquit: The Value of Forensic Science Testimony'. *Journal of Empirical Legal Studies* 8(1): 21–48.

Koehler, D. J. & James, G. (2009). 'Probability Matching in Choice Under Uncertainty: Intuition Versus Deliberation'. *Cognition* 113(1): 123–127.

Kollins, S. H. & Sparrow, E. P. (2010). *Guide to Assessment Scales in Attention-deficit/Hyperactivity Disorder.* London: Springer Healthcare.

Kondo, T., Zakany, J., Innis, W. J. & Duboule, D. (1997). 'Of Fingers, Toes, and Penises'. *Nature* 390(6655): 29.

Konrad, K. & Eickhoff, S. B. (2010). 'Is the ADHD Brain Wired Differently? A Review on Structural and Functional Connectivity in Attention Deficit Hyperactivity Disorder'. *Human Brain Mapping,* 31(6) 904–16.

Kotler, P. (1973). 'Atmospherics as a Marketing Tool'. *Journal of Retailing* 49(4): 48–64.

Koyama, M., Saji, F., Takahashi, S., Takemura, M., Samejima, Y., Kameda, T., Kimura, T. & Tanizawa, 0. (1991). 'Probabilistic Assessment of the HLA Sharing of Recurrent Spontaneous Abortion Couples in the Japanese Population'. *Tissue Antigens* 37(5): 211–17.

Kraft-Ebbing, R. (1965). *Psychopathia Sexualis.* New York: Putnam & Sons.

Kuhn, D. (2006). 'Do Cognitive Changes Accompany Developments in the Adolescent Brain?' *Perspectives on Psychological Science* 1(1): 59–67.

Kurzban, R. (2010). 'Does the Brain Consume Additional Glucose During Self-Control Tasks?' *Evolutionary Psychology* 8(2): 244–59, 246.

Laitinen, T. (1993). 'A Set of MHC Haplotypes Found Among Finnish Couples Suffering from Recurrent Spontaneous Abortions'. *American Journal of Reproductive Immunology* 29(3): 148–54.

Lakoff, G. & Johnson, M. (1999). *Philosophy in the flesh: The Embodied Mind and its Challenge to Western Thought.* New York: HarperCollins.

Lakoff, G. & Nunez, R. E. (2000) *Where Mathematics Comes From: How the Embodied Mind Brings Mathematics into Being.* New York: Basic Books.

Langer, E. (1989). *Mindfulness.* Cambridge. Mass: Da Capo Press.

Langer, E., Blank, A. & Chanowitz, B. (1978). 'The Mindlessness of Ostensibly Thoughtful Action: The Role of "Placebic" Information in Interpersonal Interaction'. *Journal of Personality and Social Psychology* 36(6), 635–42.

Lansbergen, M. M. (2007). Impulsivity: A Deficiency of Inhibitory Control? *Thesis, Utrecht University.*

Lansbergen, M. M., Schutter, D. J. L. G. & Kenemans, L. (2007).

'Subjective Impulsivity and Baseline EEG in Relation to Stopping Performance'. *Brain Research* 1148: 161–9.

Lansbergen, M. M., Van-Dongen-Boomsma, M., Buitelaar, J. K. & Slaats-Willemse, D. (2010). 'ADHD and EEG-Neurofeedback: A Double-Blind Randomized Placebo-Controlled Feasibility Study'. *Journal of Neural Transmission* 118(2): 275–84.

Larner, A. & Leach, J. P. (2002). 'Phineas Gage and the Beginnings of Neuropsychology'. *ACNR* 2(3): 26

Larrick, R. P., Timmerman, T. A., Carton, A. M. & Abrevaya, J. (2011). 'Temper, Temperature, and Temptation: Heat-Related Retaliation in Baseball'. *Psychological Science* 22(4): 423–8.

Lau, H. C., Rogers, R. D. & Passingham, R. E. (2006). 'On Measuring the Perceived Onsets of Spontaneous Actions'. *Journal of Neuroscience* 26(27): 7265–71.

Lau, H. C., Rogers, R. D. & Passingham, R. E. (2007). 'Manipulating the Experienced Onset of Intention after Action Execution'. *Journal of Cognitive Neuroscience* 19(1): 81–90.

Lavater, J. C. (1880). 'Essays on Physiognomy; for the Promotion of the Knowledge and the Love of Mankind' (Gale Document Number CW114125313). Retrieved 15th May 2005 from Gale Group, *Eighteenth Century Collections Online.* (Original work published 1772).

Lavie, N. (2010). 'Attention, Distraction, and Cognitive Control under Load'. *Current Directions in Psychological Science* 19(3): 143–8.

Lavrakas, P. J. (1975). 'Female Preferences for Male Physiques'. *Paper Presented at the Meeting of the Midwestern Psychological Association, Chicago, May.*

Lebhar, G. M. (1963). *Chain Stores in America, 1859–1962* (3rd edition). New York: Chain Store Publishing Corporation 226–8.

Le Bon, G. (1895/2002). *The Crowd: A Study of the Popular Mind.* New York: Dover Publications 8.

Lee, L., Amir, O. & Ariely, D. (2009). 'In Search of Homo Economicus: Cognitive Noise and the Role of Emotion in Preference Consistency'. *Journal of Consumer Research* 36(2): 173–87.

Lee, J. & Johnson, K. K. P. (2010). 'Buying Fashion Impulsively: Environmental and Personal Influences'. *Journal of Global Fashion Marketing 1(1): 30–9.*

Lee, T-W., Wu, Y-T., Yu, Y. W-Y., Wu, H-C. & Chen, T-J. (2012). 'A Smarter Brain is Associated with Stronger Neural Interaction in Healthy Young Females: A Resting EEG Coherence Study'. *Intelligence* 40(1): 38–48.

Leotti, L. A. & Delgado, M. R. (2011). 'The Inherent Reward of Choice'. *Psychological Science* 22(10): 1310–18.

Lerner, J. S., Small, D. A. & Loewenstein, G. (2004). 'Heart Strings and Purse Strings: Carryover Effects of Emotions on Economic Decisions'. *Psychological Science* 15(5): 337–41.

Levitt, T. (1975). 'Marketing Myopia'. *Harvard Business Review*, Sept–Oct, No. 75507 (reprint).

Lewis, D. (1978). *The Secret Language of Your Child.* London: Souvenir Press.

Lewis, D. (1985). *Loving and Loathing.* London: Constable & Co.

Lewis, D. (1997). *Dying for Information? An Investigation into the Effects of Information Overload in the UK and Worldwide.* London: Reuters Business Information.

Lewis, D. (1999). *Information Overload. Practical Strategies for Surviving in Today's Workplace.* Harmondsworth: Penguin.

Lewis, D. (2012). 'Retail Atmospherics: A Practical Guide to Serving Your Customers Right in the 21st Century'. Free download from www.themindlab.org.

Lewis, D. & Greene, R. (1982). *Thinking Better.* New York: Rawson Wade.

Lewis, D. & Bridger, D. (2001). *The Soul of the New Consumer: Authenticity – What We Buy and Why In the New Economy.* London: Nicholas Brealey Publishing.

Lewis, M. D. (2011). 'Dopamine and the Neural "Now"'. *Perspectives on Psychological Science* 6(2): 150–5.

Lhermitte, F. (1983). '"Utilisation Behaviour" and its Relation to Lesions of the Frontal Lobes'. *Brain* 106(2): 237–55.

Lhermitte, F., Pillon, B. & Serdaru, M. (1986). 'Human Autonomy and the Frontal Lobes. Part I: Imitation and Utilization Behavior: A Neuropsychological Study of 75 Patients'. *Annals of Neurology* 19(4): 326–34.

Li, W., Moallem, I., Paller, K. A. & Gottfried, J. A. (2007). 'Subliminal

Smells Can Guide Social Preferences'. *Psychological Science* 18(12): 1044–9.

Libet, B., Gleason, C. A., Wright, E. W. & Pearl D. K. (1983). 'Time of Conscious Intention to Act In Relation to Onset Of Cerebral Activity (Readiness Potential): The Unconscious Initiation of a Freely Voluntary Act'. *Brain* 106(Pt3): 623–42.

Liebowitz, M. R. (1983). *The Chemistry of Love.* New York: Little Brown.

Logan, G. D., Schachar, R. J. & Tannock, R. (1997). 'Impulsivity and Inhibitory Control'. *Psychological Science* 8(1): 60–4.

Lowe, M. R., Van Steenburgh, J., Ochner, C. & Coletta, M. (2009). 'Neural Correlates of Individual Differences Related to Appetite'. *Physiology & Behaviour* 97(5): 561–71.

Luchins, A. & Luchins, E. (1942). 'Mechanisation in Problem-Solving: The Effect of Einstellung'. *Psychological Monographs* 54(6): 52–3.

Ma, Y., Wang, C. & Han, S. (2011). 'Neural Responses to Perceived Pain in Others Predict Real-Life Monetary Donations in Different Socioeconomic Contexts'. *Neuroimage* 57(3): 1273–80.

MacFarlane, A. (1975). 'Olfaction in the Development of Social Preferences in the Human Neonate'. *Ciba Foundation Symposium* 33: 103–17.

Macmillan, M. (2000). *An Odd Kind of Fame. Stories of Phineas Gage.* Cambridge, Mass: The MIT Press.

Macmillan, M. (2008). 'Phineas Gage – Unravelling the Myth'. *The Psychologist* 21(9): 828–31.

Maddux, W. W., Yang, H., Falk, C., Adam, H., Adair, W., Endo, Y., Carmon, Z. & Heine, S. J. (2010). 'For Whom Is Parting With Possessions More Painful? Cultural Differences In the Endowment Effect'. *Psychological Science* 21(12): 1910–17.

Madey, S. F., Simo, M., Dillworth, D., Kemper, D., Toczynski, A. & Perella, A. (1996). 'They Do Get More Attractive at Closing Time, But Only When You Are Not In a Relationship'. *Basic and Applied Social Psychology* 18(4): 387–93.

Madhavaram, S. R. & Laverie, D. A. (2004). 'Exploring Impulse Purchasing on the Internet'. *Advances in Consumer Research* 31(1): 59–66.

Maffezioli, P. (2009). 'Commentary of Micromotives and Macrobehavior by T. C. Schelling'. *Humana. Mente* 10: 199–206.

Maine de Biran, F. P. G. (1803/1929). *The Influence of Habit on the Faculty of Thinking*, (Boehn, M. D., Trans.). London: Baillière & Co.

Malkani, G. (2011). 'Britain Burns the Colour of "*A Clockwork Orange*"'. *Financial Times* 12th August: 9.

Manning, J. T. (2002). *Digit Ratio: A Pointer to Fertility, Behaviour, and Health*. New Brunswick, NJ: Rutgers University Press.

Manning, J. T., Scott, D., Wilson, J. & Lewis-Jones, D. I. (1998). 'The Ratio of 2nd to 4th Digit Length: a Predictor of Sperm Numbers and Concentration of Testosterone, Luteinizing Hormone and Oestrogen'. *Human Reproduction* 13(11): 3000–3004.

Mantonakis, A., Rodero, P., Lesschaeve, I. & Hastie, R. (2009). 'Order in Choice: Effects of Serial Position on Preferences'. *Psychological Science* 20(11): 1309–12.

Marlowe, F., Westman, A. (2001). 'Preferred Waist-to-Hip Ratio and Ecology'. *Personality and Individual Differences* 30(3): 481–9.

Marlowe, F., Apicella, C. & Reed, D. (2005). 'Men's Preferences for Women's Profile Waist-to-Hip Ratio in Two Societies'. *Evolution and Human Behavior* 26(6): 458–68.

Martineau, H. (1983). *Autobiography*. (Weiner, G., Ed.), Vol. 2, London: Virago175–7.

Masicampo, E. J. & Baumeister, R. F. (2008). 'Toward a Physiology of Dual-Process Reasoning and Judgment: Lemonade, Willpower, and Expensive Rule-Based Analysis'. *Psychological Science* 19(3): 255–60.

Masuda, T. (2006). *Culture and Attention: Comparing Cultural Variations in Patterns of Eye-Movements between East Asians and North Americans*. Presentation at the 5th International Conference of the Cognitive Science, Vancouver, Canada.

Masuda, T. & Nisbett, R. E. (2001). 'Attending Holistically Versus Analytically: Comparing the Context Sensitivity of Japanese and Americans'. *Journal of Personality and Social Psychology* 81(5): 922–34.

Mataró, M., Jurado, M. A., García-Sánchez, C., Barraquer, L., Costa-Jussá, F. R. & Junqué, C. (2001). 'Long-Term Effects of Bilateral Frontal Brain Lesion 60 Years After Injury With an Iron Bar'. *Archives of Neurology* 58(7): 1139–42.

Mather, M., Gorlick, M. A. & Lighthall, N. R. (2009). 'To Brake or

Accelerate When the Light Turns Yellow? Stress Reduces Older Adults' Risk Taking in a Driving Game'. *Psychological Science* 20(2): 174–6.

Matier, P. & Ross, A. (2005). 'Film Captures Suicides on Golden Gate Bridge'. *San Francisco Chronicle* 19th January.

Matsunaga, H., Kiriike, N., Iwasaki, Y., Matsui, T., Nagata, T., Yamagami, S. & Kaye, W. H. (2000). 'Multi-Impulsivity among Bulimic Patients in Japan'. *The International Journal of Eating Disorders* 27(3): 348–52.

Matthews, R. (1997). 'How Right Can You Be?' *New Scientist* No.2072: 28–31, 8th March.

Mazur, A. (1986). 'U.S. Trends in Feminine Beauty and Over adaptation'. *Journal of Sex Research* 22: 281–303.

McAnarney, E. R. (2008). 'Adolescent Brain Development: Forging New Links?' *Journal of Adolescent Health* 42(4): 321–3.

McClintock, M. K. (1971). 'Menstrual Synchrony and Suppression'. *Nature* 229(5282): 244–5.

McClure, R. F. (1986). 'Self Control and Achievement Motivation in Young and Old Subjects'. *Psychology: A Journal of Human Behaviour* 23(1): 20–2.

McCormick, P. A. (1997). 'Orienting Attention without Awareness'. *Journal of Experimental Psychology. Human Perception and Performance* 23(1): 168–80.

McCown, W. (1993). 'The Ideodynamics of Impulsive Families'. *Paper presented at the 101st Annual Convention of the American Psychological Association, Toronto, Canada.*

McCown, W. G. & DeSimone, P. A. (1993). 'Impulses, Impulsivity, and Impulsive Behaviours: a Historical Review of a Contemporary Issue'. In McCown, W. G., Johnson, J., Shure, M. B., (Eds.) *The Impulsive Client: Theory, Research, and Treatment.* Washington, DC: American Psychological Association. 8.

McIntosh, J. L. (2010). 'U.S.A. Suicide: 2007 Official Final Data'. *American Association of Suicidology.*

McIntyre, M. H., Barrett, E. S., McDermott, R., Johnson, D. D. P., Cowden, J. & Rosen, S. P. (2007). 'Finger Length Ratio (2D: 4D) and Sex Differences in Aggression During a Simulated War Game'. *Personality and Individual Differences* 42(4): 755–64.

McNees, M. P., Egli, D. S., Marshall, R. S., Schnelle, J. F. & Risley, T. R.

(1976). 'Shoplifting Prevention: Providing Information Through Signs'. *Journal of Applied Behavior Analysis* 9(4) 399–405.

McNeil, B. J., Pauker, S. G., Sox, H. C. Jr. & Tversky, A. (1982). 'On the Elicitation of Preferences for Alternative Therapies'. *New English Journal of Medicine* 306(21): 1259–62.

Mead, N. L., Baumeister, R. F., Gino, F., Schwitzer, M. E. & Ariely, D. (2009). 'Too Tired to Tell the Truth: Self-Control Resource Depletion and Dishonesty'. *Journal of Experimental Social Psychology* 45(3): 594–7.

Mesoudi, A. (2009). 'The Cultural Dynamics of Copycat Suicide'. *PLoS One* 4(9): e7252.

Meyer-Lindenberg, A., Buckholtz, J. W., Kolachana, B., Hariri, A. R., Pezawas, L., Blasi, G., Wabnitz, A., Honea, R., Verchinski, B., Callicott, J. H., Egan, M., Mattay, V. & Weinberger, D. R. (2006). 'Neural Mechanisms of Genetic Risk for Impulsivity and Violence in Humans'. *PNAS* 103(16): 6269–74.

Meyers, C. A., Berman, S. A., Scheibel, R. S. & Hayman, A. (1992). 'Case Report: Acquired Antisocial Personality Disorder Associated with Unilateral Left Orbital Frontal Lobe Damage'. *Journal of Psychiatric Neuroscience* 17(3): 121–5.

Michalczuk, R., Bowden-Jones, H., Verdejo-Garcia, A. & Clark, L. (2011). 'Impulsivity and Cognitive Distortions in Pathological Gamblers Attending the UK National Problem Gambling Clinic: A Preliminary Report'. *Psychological Medicine* (online publication).

Miellet, S., Zhou, X., He, L., Rodger, H. & Caldara, R. (2010). 'Investigating Cultural Diversity for Extrafoveal Information Use in Visual Scenes'. *Journal of Vision* 10(6):21: 1–18.

Mikulak, A. (2011). 'Is There the Courage to Change America's Diet?' *Observer APA* 24(6): 15–17.

Milkman, K. L., Rogers, T. & Bazerman, M. H. (2008). 'Harnessing Our Inner Angels and Demons: What We Have Learned About Want/Should Conflicts and How that Knowledge Can Help Us Reduce Short-Sighted Decision Making'. *Perspectives on Psychological Science* 3(4): 324–38.

Milkman, K. L., Chugh, D. & Bzerman, M. H. (2009). 'How Can Decision Making be Improved?' *Perspectives on Psychological Science* 4(4): 379–83.

Miller, D. (1998). *Shopping, Place, and Identity.* New York: Routledge.

Miller, D. (2005). *A Theory of Shopping.* Cambridge: Polity Press.

Miller, G. (2011). 'Blue Brain Founder Responds to Critics, Clarifies His Goals'. *Science,* 334(6057): 748–9.

Miller, G. E., Lachman, M. E., Chen, E., Gruenewald, T. L., Karlamangla, A. S., Seeman & T. E. (2011). 'Pathways to Resilience: Maternal Nurturance as a Buffer Against the Effects of Childhood Poverty on Metabolic Syndrome at Midlife'. *Psychological Science* 22(12): 1591–9.

Miller, H. C., Pattison, K. F., DeWall, C. N., Rayburn-Reeves, R. & Zentall, T. R. (2010). 'Self-Control Without a "Self"? Common Self-Control Processes in Humans and Dogs'. *Psychological Science* 21(4): 534–8.

Minsky, M. (1986). *The Society of Mind.* New York: Simon and Schuster. 339.

Mischel, H. N. & Mischel, W. (1983). 'The Development of Children's Knowledge of Self-Control Strategies'. *Child Development* 54(3): 603–19.

Mischel, W., Shoda, Y., Rodriguez, M. I. (1989). 'Delay of Gratification in Children'. *Science* 244(4907): 933–8.

Mishra, H., Mishra, A. & Shiv, B. (2011). 'In Praise of Vagueness: Malleability of Vague Information as a Performance Booster'. *Psychological Science* 22(6): 733–8.

Mithen, S. (2002). *The Cognitive Basis of Science* (Carruthers, P., Stich, S., Siegal, M., Eds.). Cambridge: Cambridge University Press, 33–4.

Mobbs, D., Lau, H. C., Jones, O. D. & Frith, C. D. (2007). 'Law, Responsibility, and the Brain'. *PLoS Biology* 5(4): e103, 693–700.

Moffitt, T. E. (2005). 'The New Look of Behavioral Genetics in Developmental Psychopathology: Gene-Environment Interplay in Antisocial Behaviors'. *Psychological Bulletin* 131(4): 533–54.

Møller, A. P., Soler, M. & Thornhill, R. (1995). 'Breast Asymmetry, Sexual Selection and Human Reproductive Success'. *Ethology and Sociobiology* 16(3): 207–19.

Molnar-Szakacs, I. & Overy, K. (2006). 'Music and Mirror Neurons: From Motion to 'E' Motion'. 1(3). 235–41.

Montoya, R. M. (2007). 'Gender Similarities and Differences in Preferences for Specific Body Parts'. *Current Research in Social Psychology* 13(11): 133–44.

Morgan, K., Holmes, T. M., Schlaut, J., Marchuk, L., Kovithavongs, T., Pazderka, F. & Dossetor, J. B. (1980). 'Genetic Variability of HLA in the Dariusleut Hutterites. A Comparative Genetic Analysis of the Hutterities, the Amish, and Other Selected Caucasian Populations'. *American Journal of Human Genetics* 32(2): 246–57.

Morison, S. E. (1965). *The Oxford History of the American People.* Oxford: Oxford University Press.

Morris, J. A. (2011). 'The Conscious Mind and its Emergent Properties; an Analysis Based on Decision Theory'. *Medical Hypotheses* 77(2): 253–7.

Morrison, M. (2002). The Power of In-Store Music and its Influence on International Retail Brands and Shopper Behaviour: A Multi-Case Study Approach.

Morse, S. (2006). 'Brain Overclaim Syndrome and Criminal Responsibility: A Diagnostic Note'. *Ohio State Journal of Criminal Law* 3: 397–412.

Most, S. B., Simons, D. J., Scholl, B. J. & Chabris, C. F. (2000). 'Sustained Inattentional Blindness: The Role of Location in the Detection of Unexpected Dynamic Events'. *Psyche* 6(14).

Muraven, M. (2010). 'Practicing Self-Control Lowers the Risk of Smoking Lapse'. *Psychology of Addictive Behaviors* 24(3): 446–52.

Muraven, M., Tice, D. M. & Baumeister, R. F. (1998). 'Self-Control as Limited Resource: Regulatory Depletion Patterns'. *Journal of Personality and Social Psychology* 74(3): 774–89.

Muraven, M. & Baumeister, R. F. (2000). 'Self-Regulation and Depletion of Limited Resources: Does Self-Control Resemble a Muscle?' *Psychological Bulletin* 126(2): 247–59.

Muraven, M., Lorraine, C. R. & Neinhaus, K. (2002). 'Self-Control and Alcohol Restraint: An Initial Application of the Self-Control Strength Model'. *Psychology of Addictive Behaviours* 16(2): 11320.

Murphy, E. R., Illes, J. & Reiner, P. B. (2008). 'Neuroethics of Neuromarketing'. *Journal of Consumer Behaviour* 7(4–5): 293–302.

Mussweiler, T., Strack, F. & Pfeiffer, T. (2000). 'Overcoming the Inevitable Anchoring Effect: Considering the Opposite Compensates for Selective Accessibility'. *Personality and Social Psychology Bulletin* 26(9): 1142–50.

Mussweiler, T., Englich, B. & Strack, F. (2004). 'Anchoring Effect'. In

Pohl, R., (Ed.) *Cognitive Illusions – A Handbook on Fallacies and Biases in Thinking, Judgement, and Memory.* London: Psychology Press.

Myers, D. G. (2007). 'The Powers and Perils of Intuition, Understanding the Nature of our Gut Instinct'. *Scientific American Mind* 18(3): 24–31.

Myrseth, K. O. R., Fishbach, A. & Trope, Y. (2009). 'Counteractive Self-Control'. *Psychological Science* 20(2): 159–63.

Naisbitt, J. (1984). *Megatrends.* London: MacDonald and Co.

Nass, C. & Moon, Y. (2000). 'Machines and Mindlessness: Social Responses to Computers'. *Journal of Social Issues* 56(1): 81–103.

Nederkoorn, C., Smulders, F. T., Havermans, R. C., Roefs, A. & Jansen, A. (2006a). 'Impulsivity in Obese Women'. *Appetite* 47(2) 253–6.

Nederkoorn, C., Jansen, E., Mulkens, S. & Jansen, A. (2006b). 'Impulsivity Predicts Treatment Outcome in Obese Children'. *Behaviour Research and Therapy* 45(5): 1071–5.

Nederkoorn, C., Braet, C., Van Eijs, Y., Tanghe, A. & Jansen, A. (2006c). 'Why Obese Children Cannot Resist Food: The Role of Impulsivity'. *Eating Behaviors* 7(4): 315–22.

Neys, W. D. (2006). 'Dual Processing in Reasoning: Two Systems but One Reasoner'. *Psychological Science* 17(5): 428–33.

Niederkrotenthaler, T., Voracek, M., Herberth, A., Till, B., Strauss, M., Etzersdorfer, E., Eisenwort, B. & Sonneck, G. (2010). 'Role of Media Reports in Completed and Prevented Suicide: Werther v. Papageno Effects'. *The British Journal of Psychiatry* 197(3): 234–43.

Nordgren, L. F., Harreveld, F. V. & Pligt, J. V. D. (2009). 'The Restraint Bias: How the Illusion of Self-Restraint Promotes Impulsive Behavior'. *Psychological Science* 20(12): 1523–8.

Nordgren, L. F. & Chou, E. Y. (2011). 'The Push and Pull of Temptation: The Bidirectional Influence of Temptation on Self-Control'. *Psychological Science* 22(11): 1386–90.

Norretranders, T. (1998). *The User Illusion.* New York: Penguin Putnam Inc.

Ober, C., Weitkamp, L. R., Cox, N., Dytch, H., Kostyu, D. & Elias, S. (1997). 'HLA and Mate Choice in Humans'. *American Journal of Human Genetics* 61(3): 497–504.

O'Brien, B. J. (1989). 'Words or Numbers? The Evaluation of Probability Expressions in General Practice'. *Journal of the Royal College of General Practitioners* 39(320): 98–100.

O'Brien, L. (1991). *Retailing: Shopping, Society, Space*. London: David Fulton Publishers.

O'Brien, M. & Kellan, A. (2011). 'Science of Shopping: Cameras and Software That Track Our Shopping Behaviour'. *Science Nation* 1st August.

Olson, S. L., Bates, J. E. & Bayles, K. (1990). 'Early Antecedents of Childhood Impulsivity: The Role of Parent-Child Interaction, Cognitive Competence, and Temperament'. *Journal of Abnormal Child Psychology* 18(3): 176–83.

Oppenheimer, D. M. (2003). 'Not So Fast! (And Not So Frugal!): Rethinking the Recognition Heuristic'. *Cognition* 90(1): B1–9.

Ostrovsky, Y., Andalman, A. & Sinha, P. (2006). 'Vision Following Extended Congenital Blindness'. *Psychological Science* 17(12): 1009–14.

Ostrovsky, Y., Meyers, E., Ganesh, S., Mathur, U. & Sinha, P. (2009). 'Visual Parsing After Recovery from Blindness'. *Psychological Science* 20(12): 1484–91.

Patoine, B. (2009). 'The Chemistry of Love: In Search of the Elusive Human Pheromones'. *Briefing Paper for The Dana Foundation, New York*.

Patrick, C. J., Hicks, B. M., Krueger, R. F. & Lang, A. R. (2005). 'Relations Between Psychopathy Facets and Externalizing in a Criminal Offender Sample'. *Journal of Personality Disorders* 19(4): 339–56.

Patton, J. H., Stanford, M. S. & Barratt, E. S. (1995). 'Factor Structure of the Barratt Impulsiveness Scale'. *Journal of Clinical Psychology* 51(6): 768–74.

Payne, J. W., Samper, A., Bettman, J. R. & Luce, M. F. (2008). 'Boundary Conditions on Unconscious Thought In Complex Decision Making'. *Psychological Science* 19(11): 1118–23.

Pelchat, M. L. (2009). 'Food Addiction: Fact or Fiction?' *The Journal of Nutrition* 139(3): 620–2.

Pelchat, M. L., Johnson, A., Chan, R., Valdex, J. & Ragland, J. D. (2004). 'Images of Desire: Food-Craving Activation During fMRI'. *Neuroimage* 23(4): 1486–93.

Penfield, W. & Rasmussen, T. (1950). *The Cerebral Cortex of Man: A Clinical Study of Localization of Function*. The Macmillan Company.

Pennebaker, J. W., Dyer, M. A., Caulkins, R. S., Litowitz, D. L., Ackerman, P. L., Anderson, D. B. & McGraw, K. M. (1979). 'Don't the Girls Get Prettier at Closing Time: A Country and Western Application to Psychology'. *Personality and Social Psychology Bulletin* 5(1): 122–5.

Petersen, J. L. & Hyde, J. S. (2010). 'A Meta-Analytic Review of Research on Gender Differences in Sexuality, 1993–2007'. *Psychological Bulletin* 136(1): 21–38.

Petrovich, G. D., Holland, P. C. & Gallagher, M. (2005). 'Amygdala and Prefrontal Pathways to the Lateral Hypothalamus are Activated by a Learned Cue That Stimulates Eating'. *Journal of Neuroscience* 25(36): 8295–8302.

Pickering A. D. & Gray J. A. (1999). 'The Neuroscience of Personality'. In Pervin, L. A., John, O. P., (Eds.) *Handbook of Personality: Theory and Research.* 2nd edn. New York: Guilford Press 277–99.

Pierce, J. L., Kostova, T., Dirks, K. T. (2003). 'The State of Psychological Ownership: Integrating and Extending A Century of Research'. *Review of General Psychology* 7(1): 84–107.

Pisella, L., Gréa, H., Tilikete, C., Vighetto, A., Desmurget, M., Rode, G., Boisson, D. & Rossetti, Y. (2000). 'An "Automatic Pilot" for the Hand in Human Posterior Parietal Cortex: Toward Reinterpreting Optic Ataxia'. *Nature Neuroscience* 3(7): 729–36.

Plato, *The Republic* (Cornford, F. M., Trans.). Oxford: Oxford University Press.

Plous, S. (1989). 'Thinking the Unthinkable: The Effects of Anchoring on Likelihood Estimates of Nuclear War'. *Journal of Applied Social Psychology* 19(1): 67–91.

Polanczyk, G., de Lima, M. S., Horta, B. L., Biederman, J. & Rohde L. A. (2007). 'The Worldwide Prevalence of ADHD: A Systematic Review and Metaregression Analysis'. *American Journal of Psychiatry* 164(6): 942–8.

Posner, M. & Snyder, C. R. R. (1975). 'Attention and Cognitive Control'. In Soloso, R. L., (Ed.) *Information Processing and Cognition: The Loyola Symposium.* New York: Wiley, 55–8.

Postmes, T., Spears, R. & Lea, M. (2000). 'The Formation of Group Norms in Computer-Mediated Communication'. *Human Communication Research* 26(3): 341–71.

Prelec, D. & Loewenstein, G. (1991). 'Decision Making Over Time and

Under Uncertainty: A Common Approach'. *Management Science* 37(7): 770–86.

Proulx, T. & Heine, S. J. (2009). 'Connections from Kafka: Exposure to Meaning Threats Improves Implicit Learning of an Artificial Grammar'. *Psychological Science* 20(9): 1125–31.

Putz, D. A., Gaulin, S. J. C., Sporter, R. J. & McBurney, D. H. (2004). 'Sex Hormones and Finger Length: What does 2D: 4D indicate?' *Evolution and Human Behavior* 25(3): 182–99.

Pyszczynski, T., Greenberg, J. & Solomon, S. (1999). 'A Dual-Process Model of Defence Against Conscious and Unconscious Death-Related Thoughts An Extension of Terror Management Theory'. *Psychological Review* 106(4): 835–45.

Rachlin, H. (1990). 'Why Do People Gamble and Keep Gambling Despite Heavy Losses?' *Psychological Science* 1(5): 294–7.

Radovic, S. (1998). 'The Controlling Soul and the Automatic Body – A Critical Account of the Control-Automaticity Distinction'. Poster at *Toward a Science of Consciousness, Tucson III.*

Raine, A. (1993). *The Psychopathology of Crime: Criminal Behaviour As a Clinical Disorder.* San Diego: Academic Press 377.

Ratiu, P., Talos, I. F., Haker, S., Liberman, D. & Everett, P. (2004). 'The Tale of Phineas Gage, Digitally Remastered'. *Journal of Neurotrauma* 21(5): 637–43.

Reber, A. S. (1993). *Implicit Learning and Tacit Knowledge.* Oxford: University Press.

Reicher, S., Haslam, S. A. & Rath, R. (2008). 'Making a Virtue of Evil: A Five-Step Social Identity Model of the Development of Collective Hate'. *Social and Personality Psychology Compass* 2(3): 1313–44.

Reicher, S. & Stott, C. (2012). *Myths and Realities of the 2011 Riots.* London: Constable & Robinson Ltd.

Rhodes, G. (2006). 'The Evolutionary Psychology of Facial Beauty'. *Annual Review of Psychology* 57: 199–226.

Robbins, T. W., Gillan, C. M., Smith, D. G., de Wit, S. & Ersche, K. D. (2012). 'Neurocognitive Endophenotypes of Impulsivity and Compulsivity: Towards Dimensional Psychiatry'. *Trends in Cognitive Sciences* 16(1): 81–91.

Roberts, S. A., Simpson, D. M., Armstrong, S. D., Davidson, A. J.,

Robertson, D. H., McLean, L., Beynon, R. J. & Hurst, J. L. (2010). 'Darcin: A Male Pheromone that Stimulates Female Memory and Sexual Attraction to an Individual Male's Odour'. *BMC Biology* 8(1): 75

Roiser, J. P., de Martino, B., Tan, G. C. Y., Kumaran, D., Seymour, B., Wood, N. W. & Dolan, R. J. (2009). 'A Genetically Mediated Bias in Decision Making Driven by Failure of Amygdala Control'. *The Journal of Neuroscience* 29(18): 5985–91.

Rolls, E. T., Grabenhorst, F. & Parris, B. A. (2008). 'Warm Pleasant Feelings in the Brain'. *Neuroimage* 41(4): 1504–13.

Rook, D. W. & Hoch, S. J. (1985). 'Consuming Impulses'. *Advances in Consumer Research* 12(1): 23–7.

Rook, D. W. (1987). 'The Buying Impulse'. *The Journal of Consumer Research* 14(2): 189–99.

Rosburg, T., Mecklinger, A. & Frings, C. (2011). 'When the Brain Decides: A Familiarity-Based Approach to the Recognition Heuristic as Evidenced by Event-Related Brain Potentials'. *Psychological Science* 22(12): 1527–34.

Rose, C. (2011). 'The Relevance of Darwinian Selection to an Understanding of Visual Art'. *Unpublished Paper.*

Rossiter, T. (2004) 'The Effectiveness of Neurofeedback and Stimulant Drugs in Treating AD/HD: Part I. Review of Methodological Issues'. *Applied Psychophysiology Biofeedback* 29(2): 95–112.

Rossiter, T. (2004). 'The Effectiveness of Neurofeedback and Stimulant Drugs in Treating AD/HD: Part II'. Replication. *Applied Psychophysiology Biofeedback* 29(4); 233–43.

Rothbart, M. K. (1988). 'Temperament and the Development of Inhibited Approach'. *Child Development*, 59(5): 1241–50.

Rowe, A. D., Bullock, P. R., Polkey, C. E. & Morris, R. G. (2001). 'Theory of Mind' Impairments and Their Relationship to Executive Functioning Following Frontal Lobe Excisions. *Brain* 124(3): 600–616.

Rozin, P., Millman, L. & Nemeroff, C. (1986). 'Operation of the Laws of Sympathetic Magic in Disgust and Other Domains'. *Journal of Personality and Social Psychology* 50(4): 703–12.

Russell, B. (1930). *The Conquest of Happiness.* New York: W.W. Norton & Co.

Russell, B. (1947). *History of Western Philosophy.* London: George Allen and Unwin Ltd, 53.

Russell, R. (2003). 'Sex, Beauty, and the Relative Luminance of Facial Features'. *Perception,* 32(9): 1093–1107.

Russell, R. (2009): 'A Sex Difference in Facial Contrast and its Exaggeration by Cosmetics'. *Perception* 38(8): 1211–219.

Ryle, A. (1979). 'The Focus in the Brief Interpretative Psychotherapy: Dilemmas, Traps and Snags as Target Problems'. *British Journal of Psychiatry* 134(1): 46–54.

Saad, G., Nepomuceno, M. V. & Mendenhall, Z. (2011). 'Testosterone and Domain-Specific Risk: Digit Ratios (2D:4D and rel2) as Predictors of Recreational, Financial, and Social Risk-Taking Behaviours'. *Personality and Individual Differences* 51(4): 412–16.

Sabri, M., Radnovich, A. J., Li, T. Q. & Kareken, D. A. (2005). 'Neural Correlates of Olfactory Change Detection'. *Neuroimage* 25(3): 969–74.

Sandman, C. A., Davis, E. P. & Glynn, L. M. (2012). 'Prescient Human Fetuses Thrive'. *Psychological Science* 23(1) 93–100.

Sartre, J. P. (1943/1969). *Being and Nothingness: A Phenomenological Essay on Ontology.* New York: Philosophical Library. London: Melhuen & Co, Ltd 591–592.

Schiffman, L. G. & Kanuk, L. L. (2004). *Consumer Behavior.* Upper Saddle River, NJ: Pearson Prentice Hall.

Schiller, D. (2011). 'Affective Neuroscience: Tracing the Trace of Fear'. *Current Biology,* 21(18): R695–R696.

Schlosser, E. (2001). *Fast Food Nation.* New York: Houghton Mifflin Co.

Schopenhauer, A. (1851/1970). *Essays and Aphorisms* (Hollingdale, R. J Trans.). London: Penguin Books.

Schorn, R., Maurhart, B. (2009). 'Influencing Willingness to Pay by Supraliminally Priming the Concept of Honesty'. *Advances in Consumer Research* 36(1): 463–6.

Schwarz, N., Bless, H., Strack, F., Klumpp, G., Rittenaure-Schatka, H. & Simons, A. (1991). 'Ease of Retrieval as Information: Another Look at the Availability Heuristic'. *Journal of Personality and Social Psychology* 61(2): 195–202.

Schwarz, N. & Clore, G. L. (2003). 'Mood as Information: 20 Years Later'. *Psychological Inquiry* 14(3&4): 296–303.

Segall, M. H., Campbell, D. T. & Herskovit, M. J. (1966). *The Influence of Culture on Visual Perception*. Indianapolis: Bobbs-Merrill Co.

Sellers, P. (2002). 'Something To Prove Bob Nardelli was Stunned When Bob Mardelli Told Him He'd Never Run GE. "I Want an Autopsy!" He Demanded'. *Fortune*, (24th June).

Shah, A. K. & Oppenheimer, D. M. (2009). 'The Path of Least Resistance'. *Current Directions in Psychological Science* 18(4): 232–6.

Shamosh, N. A., DeYoung, C. G., Green, A. E., Reis, D. L., Johnson, M. R., Conway, A. R. A., Engle, R. W., Braver, T. S. & Gray, J. R. (2008). 'Individual Differences in Delay Discounting: Relation to Intelligence, Working Memory, and Anterior Prefrontal Cortex'. *Psychological Science* 19(9): 904–11.

Sharot, T., Velasquez, C. M. & Dolan, R. J. (2010). 'Do Decisions Shape Preference? Evidence from Blind Choice'. *Psychological Science* 21(9): 1231–5.

Shaw, K. (2004). *Book of Oddballs and Eccentrics*. Book Sales 463.

Shepard, R. N. (1967). 'Recognition Memory for Words, Sentences and Pictures'. *Journal of Verbal Learning and Verbal Behaviour* 6(1): 156–63.

Silver, A. L. S. (1996). 'William James and Gertrude Stein: Psychology Affecting Literature'. *Journal of the American Academy of Psychoanalysis* 24(2): 321–39.

Silverman, I. W. & Ragusa, D. M. (1990): 'Child and Maternal Correlates of Impulse Control in 24-Month-Children'. *Genetic, Social and General Psychology Monographs* 116(4): 435–73.

Simons, D. J. & Chabris, C. F. (1999). 'Gorillas in Our Midst: Sustained Inattentional Blindness for Dynamic Events'. *Perception* 28(9): 1059–74.

Simons, D. J. & Chabris, C. F., Schnur, T., Levin, D. T. (2002). 'Evidence for Preserved Representations in Change Blindness'. *Consciousness and Cognition* 11(1): 78–97.

Simonsohn, U. (2009). 'Direct Risk Aversion: Evidence from Risky Prospects Valued Below their Worst Outcome'. *Psychological Science* 20(6): 686–92.

Singh, D. (1993). 'Adaptive Significance of Female Physical Attractiveness: Role of Waist-to-Hip Ratio'. *Journal of Personality and Social Psychology* 65(2): 293–307.

Sloman, S. A. (1996). 'The Empirical Case for Two Systems of Reasoning'. *Psychological Bulletin* 119(1): 3–22.

Smilowitz, J. T., German, J. B. & Zivkovic, A. M. (2010). 'Food Intake and Obesity: The Case of Fat'. In Montmayeur J. P., le Coutre, J., (Eds.) *Fat Detection: Taste, Texture, and Post Ingestive Effects.* Boca Raton, FL: CRC Press, Ch22.

Smith, D. (2009). *The Wonder of the Teen Brain.* The Age, Melbourne, 30th March.

Smith, E. R. & DeCoster. J. (2000). 'Dual-Process Models in Social and Cognitive Psychology: Conceptual Integration and Links to Underlying Memory Systems'. *Personality and Social Psychology Review* 4(2): 108–31.

Smith, L. (1952). *A Dictionary of Psychiatry for the Layman.* London:Maxwell.

Smith, P. K., Dijksterhuis, A. & Wigboldus, D. H. (2008). 'Powerful People Make Good Decisions Even When They Consciously Think'. *Psychological Science* 19(12): 1258–9.

Smith, T. G. (2004). 'The McDonald's Equilibrium: Advertising, Empty Calories, and the Endogenous Determination of Dietary Preferences'. *Social Choice and Welfare* 23(3): 383–413.

Smullyan, R. M. (1978). *What Is the Name of this Book? The Riddle of Dracula and Other Logical Puzzles.* Englewood Cliffs, NJ: Prentice-Hall.

Solomons. L. & Stein, G. (1896). 'Normal Motor Automation'. *Psychological Review* 3: 492–572.

Sorensen, H. (2009). *Inside the Mind of the Shopper: The Science of Retailing.* Upper Saddle River, NJ: Pearson Prentice Hall.

Spinoza, B. (1677/1883). *The Ethics. Part II.* New York: Dover. 105.

Sprenger, J. & Kramer, H. (circa 1486/1968). *Malleus Maleficarum.* London: Folio edition. 18–19.

Stack, S. (2003). 'Media Coverage as a Risk Factor in Suicide'. *Journal of Epidemiology and Community Health* 57(4): 238–40.

Stack, S. & Gundlach, J. (1995). 'Country Music and Suicide – Individual, Indirect, and Interaction Effects: A Reply to Snipes and Maguire'. *Social Forces* 74(1) 331–5.

Standing, L. (1973). 'Learning 10,000 Pictures'. *Quarterly Journal of Experimental Psychology* 25(2): 207–22.

Stanford, M. S. & Barratt, E. S. (1992). 'Impulsivity and the Multi-Impulsive Personality Disorder'. *Personality and Individual Differences* 13(7): 831–4.

Stanovich, K. (2004). *The Robots Rebellion*, Chicago: Chicago University Press 34.

Stanovich, K. E. & West, R. F. (2003). *Evolution and the Psychology of Thinking* (Over, D. E., ed.). Psychology Press 171–230.

Stanton, S. J., Mullette-Gillman, O. A., McLaurin, R. E., Kuhn, C. M., LaBar, K. S., Platt, M. L. & Huettel, S. A. (2011). 'Low- and High-Testosterone Individuals Exhibit Decreased Aversion to Economic Risk'. *Psychological Science,* 22(4): 447–53.

Steel, Z. & Blaszczynski, A. (1998). 'Impulsivity, Personality Disorders and Pathological Gambling Severity'. *Addiction* 93(6): 895–905.

Stein, G. (1936). *The Autobiography of Alice B. Toklas.* New York: Random House 79

Steinberg, L. (2007). 'Risk Taking in Adolescence. New Perspectives from Brain and Behavioural Studies'. *Current Directions in Psychological Science* 16(2): 55–9.

Steinberg, L., Lamborn, S. D., Dornbusch, S. M. & Darling, N. (1992). 'Impact of Parenting Practices on Adolescent Achievement: Authoritative Parenting, School Involvement, and Encouragement to Succeed'. *Child Development* 63(5): 1266–81.

Steinberg, L. & Scott, E. S. (2003). 'Less Guilty by Reason of Adolescence: Developmental Immaturity, Diminished Responsibility, and the Juvenile Death Penalty'. *American Psychologist* 58(12): 1009–18.

Stern, H. (1962). 'The Significance of Impulse Buying Today'. *Journal of Marketing* 26(2): 59–60.

Stoddart, D. M. (1991). *The Scented Ape. The Biology and Culture of Human Odour.* Cambridge: Cambridge University Press.

Stone, V. E., Baron-Cohen, S. & Knight, R. T. (1998). 'Frontal Lobe Contributions to Theory of Mind'. *Journal of Cognitive Neuroscience:* 10(5) 640–56.

Stoner, J. (1968). 'Risky and Cautious Shifts in Group Decisions: The Influence of Widely Held Values'. *Journal of Experimental Social Psychology* 4: 442–59.

Stott, C. (2009). 'Crowd Psychology and Public Order Policing: An

Overview of Scientific Theory and Evidence'. *Report commissioned by the UK constabulary.*

Stott, C. & Drury, J. (2000). 'Crowds, Context and Identity: Dynamic Categorization Processes in the "Poll Tax Riot"'. *Human Relations* 53(2): 247–73.

Stott, C. & Reicher, S. (1998). 'How Conflict Escalates: The Inter-Group Dynamics of Collective Football Crowd "Violence"'. *Sociology* 32(2): 353–77.

Stott, C., Hutchinson, P. & Drury, J. (2001). '"Hooligans" Abroad Abroad? Inter-Group Dynamics, Social Identity and Participation in Collective "Disorder" at the 1998 World Cup Finals'. *British Journal of Social Psychology* 40(3): 359–84.

Strack, F. & Mussweiler, T. (1997). 'Explaining the Enigmatic Anchoring Effect: Mechanisms of Selective Accessibility'. *Journal of Personality and Social Psychology* 73(3): 437–46.

Strack, F. & Deutsch, R. (2004). 'Reflective and Impulsive Determinants of Social Behavior'. *Personality and Social Psychology Review* 8(3): 220–47.

Strahilevitz, M. A. & Loewenstein, G. (1998). 'The Effect of Ownership History on the Valuation of Objects'. *Journal of Consumer Research* 25(3): 276–89.

Strehl, U., Leins, U., Goth, G., Klinger, C., Hinterberger, T. & Birbaumer, N. (2006). 'Self-Regulation of Slow Cortical Potentials: A New Treatment for Children with Attention-Deficit/Hyperactivity Disorder'. *Pediatrics* 118(5): e1530–40.

Subcommittee on Attention-Deficit/Hyperactivity Disorder, Steering Committee on Quality Improvement and Management (2011). 'ADHD: Clinical Practice Guideline for the Diagnosis, Evaluation, and Treatment of Attention-Deficit/Hyperactivity Disorder in Children and Adolescents'. *Pediatrics,* 128(5): 1007–22.

Sugiyama, L. S. (2004). 'Is Beauty in the Context-Sensitive Adaptations of the Beholder? Shiwiar Use of Waist-to-Hip Ratio in Assessments of Female Mate Value'. *Evolution and Human Behavior* 25(1): 51–62.

Sukel, K. (2012). 'Dirty Minds: How Our Brains Influence Love, Sex, and Relationships'. New York: Free Press.

Sumner, P. & Husain, M. (2008). 'At the Edge of Consciousness:

Automatic Motor Activation and Voluntary Control'. *Neuroscientist* 14(5): 474–86.

Sumner, P., Edden, R. A. E., Bompass, A., Evans, C. J., Singh, K. D. (2010). 'More GABA, Less Distraction: A Neurochemical Predictor of Motor Decision Speed'. *Nature Neuroscience* 13(7): 825–7.

Sutherland, S. (1989). *Macmillan Dictionary of Psychology*. London: The Macmillan Press.

Sutherland, S. (1993). *Irrationality: The Enemy Within*. London: Constable.

Swanson, J. M. Sergeant, J. A., Taylor, E., Sonuga-Barke, E. J., Jensen, P. S. & Cantwell, D. P. (1998). 'Attention-Deficit Hyperactivity Disorder and Hyperkinetic Disorder'. *Lancet* 351(9100): 429–33.

Tabibnia, G., Satpute, A. B. & Lieberman, M. D. (2008). 'The Sunny Side of Fairness: Preference for Fairness Activates Reward Circuitry (and Disregarding Unfairness Activates Self-Control Circuitry)'. *Psychological Science* 19(4): 339–47.

Tamir, M. (2009). 'What do People Want to Feel and Why? Pleasure and Utility in Emotion Regulation'. *Current Directions in Psychological Science* 18(2): 101–5.

Tanaka, Y., Albert, M. L., Hara, H., Miyashita, T. & Kotani, N. (2000). Forced Hyperphasia and Environmental Dependency Syndrome. *Journal of Neurology, Neurosurgery and Psychiatry* 68(2): 224–6.

Tancredi, L. R. (2005). *Hardwired Behavior: What Neuroscience Reveals About Morality*. New York: Cambridge University Press.

Tangney, J. P., Baumeister, R. F. & Boone, A. L. (2004). 'High Self-Control Predicts Good Adjustment, Less Pathology, Better Grades, and Interpersonal Success'. *Journal of Personality* 72(2): 271–322.

Taylor, S. E. & Brown, J. D. (1988). 'Illusion and Well-Being: A Social Psychological Perspective on Mental Health'. *Psychological Bulletin* 103(2): 193–210.

Taylor, S. E. & Brown, J. D. (1994). 'Positive Illusions and Well-Being Revisited Separating Fact From Fiction'. *Psychological Bulletin* 116(1): 21–7.

Tetlock, P. E. (2003). 'Thinking the Unthinkable: Sacred Values and Taboo Cognitions'. *Trends in Cognitive Sciences* 7(7): 320–4.

Thaler, R. H. & Shefrin, H. M. (1981). 'An Economic Theory of Self-Control'. *The Journal of Political Economy* 89(2): 392–406.

Tirmizi, M. A., Rehman, K. U. & Saif, M. I. (2009). 'An Empirical Study of Consumer Impulse Buying Behavior in Local Markets'. *European Journal of Scientific Research* 28(4): 522–32.

Toch, H. (1982). 'The Disturbed Disruptive Inmate: Where Does the Bus Stop?' *Journal of Psychiatry and Law* 10(3) 327–49.

Todorov, A., Mandisodza, A. N., Goren, A. & Hall, C. C. (2005). 'Inferences of Competence From Faces Predict Election Outcomes'. *Science 308(5728): 1623–6.*

Trewavas, A. J. & Baluška, F. (2011). 'The Ubiquity of Consciousness'. *EMBO Reports* 12(12): 1221–5.

Tversky, A. & Kahneman, D. (1974). 'Judgment Under Uncertainty: Heuristics and Biases'. *Science* 185(4157): 1124–31.

Tversky, A. & Kahneman, D. (1986). 'Rational Choice and the Framing of Decisions'. *The Journal of Business* 59(4): 251–78.

Twitchell, J. B. (1999). *Lead Us Into Temptation: The Triumph of American Materialism.* New York: Columbia University Press.

Tybur, J. M., Bryan, A. D., Magnan, R. E. & Hooper, A. E. C. (2011). 'Smells Like Safe Sex: Olfactory Pathogen Primes Increase Intentions to Use Condoms'. *Psychological Science* 22(4): 478–80.

Underhill, P. (1999). *Why We Buy: The Science of Shopping.* New York: Simon & Schuster.

User Interface Engineering, What Causes Customers to Buy on Impulse? *E-Commerce White Paper*

Uttal, W. R. (2001). *The New Phrenology: The Limits of Localizing Cognitive Processes in the Brain.* Cambridge, Mass: MIT Press.

Van de Ven, N., T. Gilovich. & Zeelenberg, M. (2010). 'Delay, Doubt, and Decision: How Delaying a Choice Reduces the Appeal of (Descriptively) Normative Options'. *Psychological Science* 21(4): 568–73.

Van Dijk, E. & Van Knippenberg, D. (1998). 'Trading Wine: On the Endowment Effect, Loss Aversion, and the Comparability of Consumer Goods'. *Journal of Economic Psychology,* 19(4): 485–95.

Van Hout, G. C., Van Oudheusden, I. & Van Heck, G. L. (2004). 'Psychological Profile of the Morbidly Obese'. *Obesity Surgery* 14(5): 579–88.

Ventura, A. K. & Mennella, J. A. (2011). 'Innate and Learned Preferences for Sweet Taste during Childhood'. *Current Opinions in Clinical Nutrition and Metabolic Care* 14(4): 379–84.

Vigil-Colet, A., Morales-Vives, F. & Tous, J. (2008). 'The Relationships between Functional and Dysfunctional Impulsivity and Aggression across Different Samples'. *The Spanish Journal of Psychology* 11(2): 480–7.

Vohs, K. D. & Heatherton, T. F. (2000). 'Self-Regulatory Failure: A Resource-Depletion Approach'. *Psychological Science* 11(3): 249–54.

Vohs, K. D., Baumeister, R. F., Schmeichel, B. J., Twenge, J. M., Nelson, N. M. & Tice, D. M. (2008). 'Making Choices Impairs Subsequent Self-Control: A Limited-Resource Account of Decision Making, Self-Regulation, and Active Initiative'. *Journal of Personality and Social Psychology* 94(5): 883–98.

Vohs, K. D. & Schooler, J. W. (2008). 'The Value of Believing in Free Will: Encouraging a Belief in Determinism Increases Cheating'. *Psychological Science* 19(1): 49–54.

Volkow, N. D., Wang, G. J. & Baler, R. D. (2011). 'Reward, Dopamine and the Control of Food Intake: Implications for Obesity'. *Trends in Cognitive Sciences* 15(1): 37–46.

Volkow, N. D., Wang, G-J., Fowler, J. S., Tomasi, D. & Telang, F. (2011). 'Addiction: Beyond Dopamine Reward Circuitry'. *PNAS 108(37):* 15037–42.

Von Bartheld, C. S. (2004). 'The Terminal Nerve and its Relation with Extrabulbar "Olfactory" Projections: Lessons from Lampreys and Lungfishes'. *Microscopic Research and Technique* 65(1–2): 13–24.

Waddington, C. H. (1977). *Tools for Thought.* London: Jonathan Cape.

Wallace, J. F. & Newman, J. P. (1990). 'Differential Effects of Reward and Punishment Cues on Response Speed in Anxious and Impulsive Individuals'. *Personality and Individual Differences* 11(10), 999–1009.

Wang, G. J., Volkow, N. D., Logan, J., Pappas, N. R., Wong, C. T., Zhu, W., Netusil, N. & Fowler, J. S. (2001). 'Brain Dopamine and Obesity'. *Lancet* 357(9253): 354–7.

Wang, G. J., Volkow, N. D., Thanos, P. K. & Fowler, J. S. (2004). 'Similarity Between Obesity and Drug Addiction as Assessed by Neurofunctional Imaging: A Concept Review.' *Journal of Addictive Diseases* 23(3): 39–53.

Wang, G. J., Yang, J., Volkow, N. D., Telang, F., Ma, Y., Zhu, W., Wong, C. T., Tomasi, D., Thanos, P. K. & Fowler, J. S. (2006). 'Gastric

Stimulation in Obese Subjects Activates the Hippocampus and Other Regions Involved in Brain Reward Circuitry'. *Proceedings of the National Academy of Sciences of the U S A* 103(42): 15641–5.

Wang, L., Zhu, C., He, Y., Zang, Y., Cao, Q., Zhang, H., Zhong, Q. & Wang, Y. (2009). 'Altered Small-World Brain Functional Networks in Children with Attention-Deficit/Hyperactivity Disorder'. *Human Brain Mapping* 30(2): 63849.

Wargo, E. (2009). 'Resisting Temptation'. *Observer* APA 22(1): 10–17.

Webster, C. D., Menzies, R. J. & Jackson, M. A. (1982). *Clinical Assessment Before Trial: Legal Issues and Mental Disorder.* Toronto: Butterworths.

Wedekind, C., Seebeck, T., Bettens, F. & Paepke, A. J. (1995). 'MCH-Dependent Mate Preferences in Humans'. *Proceedings of the Royal Society of London,* 260(1359): 245–9.

Wegner, D. M. (1994). 'Ironic Processes of Mental Control'. *Psychological Review* 101(1): 34–52.

Wegner, D. M. (2002). *The Illusion of Conscious Will.* Cambridge, MA: MIT Press.

Wegner, D. M. (2004). 'Précis of the Illusion of Conscious Will'. *Behavioural and Brain Sciences* 27(5): 649–59.

Wegner, D. M. (2009). 'How to Think, Say, or Do Precisely the Worst Thing for Any Occasion'. *Science* 325, 48

Wegner, D. M., Carter III, S. R. & White, T. L. (1987). 'Paradoxical Effects of Thought Suppression'. *Journal of Personality and Social Psychology* 53(1): 5–13.

Wegner, D. M., Broome, A. & Blumberg, S. J. (1997). 'Ironic Effects of Trying to Relax Under Stress'. *Behaviour Research and Therapy* 35(1) 11–21.

Wegner, D. M., Ansfield, M. & Pilloff, D. (1998). 'The Putt and the Pendulum: Ironic Effects of the Mental Control of Action'. *Psychological Science* 9(3): 196–9.

Wegner, D. M. & Wheatley, T. (1999). 'Apparent Mental Causation: Sources of the Experience of Will'. *American Psychologist* 54(7): 480–91.

Wells, D. (1986). *The Penguin Dictionary of Curious and Interesting Numbers.* Harmondsworth, Middlesex: Penguin Books.

Whalen, P. J., Kagan, J., Cook, R. G., Davis, F. C., Hackjin, K., Polis, S., McLaren, D. G., Somerville, L. H., McLean, A. A., Maxwell, J. S. &

Johnstone, T. (2004). 'Human Amygdala Responsivity to Masked Fearful Eye Whites'. *Science* 306(5704): 2061.

Whitchurch, E. R., Wilson, T. D. & Gilbert, D. T. (2011). '"He Loves Me, He Loves Me Not...": Uncertainty Can Increase Romantic Attraction'. *Psychological Science* 22(2): 172–5.

Whiteside, S. P. & Lynam, D. R. (2001). 'The Five Factor Model and Impulsivity: Using a Structural Model of Personality to Understand Impulsivity'. *Personality and Individual Differences* 30(4): 669–89.

Whitson J. A. & Galinsky, A. D. (2008). 'Lacking Control Increases Illusory Pattern Perception'. *Science* 322(5898): 115–17.

Widom, C. S. (1989). 'The Cycle of Violence'. *Science* 244(4901): 160–6.

Wiggins, J. S., Wiggins, N. & Conger, J. C. (1968). 'Correlates of Heterosexual Somatic Preference'. *Journal of Personality and Social Psychology* 10(1): 82–90.

Williams, J. H., Whiten, A., Suddendorf, T. & Perrett, D. I. (2001). 'Imitation, Mirror Neurons and Autism'. *Neuroscience and Biobehavioral Reviews* 25(4): 287–95.

Willis, J. & Todorov, A. (2006). 'First Impressions: Making Up Your Mind After a 100-Ms Exposure to a Face'. *Psychological Science* 17(7): 592–8.

Wilson, T. D. & Schooler, J. W. (1991). 'Thinking Too Much: Introspection Can Reduce the Quality of Preferences and Decisions'. *Journal of Personality and Social Psychology,* 60(2): 181–92.

Winkel, D. E., Wyland, R. L., Shaffer, M. A., Clason, P. Et al. (2011). 'A New Perspective on Psychological Resources: Unanticipated Consequences of Impulsivity and Emotional Intelligence'. *Journal of Occupational and Organizational Psychology* 84(1): 78–94.

Winkielman, P., Knutson, B., Paulus, M. & Trijillo, J. L. (2007). 'Affective Influence on Judgments and Decisions: Moving Towards Core Mechanisms'. *Review of General Psychology* 11(2): 179.

Wishnie, H. A. (1977). *The Impulsive Personality: Understanding People with Destructive Character Disorders.* New York: Plenum Press. 44.

Wolf, M. J. (1999). *The Entertainment Economy.* New York: Random House. 4.

Woloshin, S. & Schwartz, L. M. (1999). 'How Can We Help People Make Sense of Medical Data?' *Effective Clinical Practice* 2(4): 176–83.

Wong, E. M., Ormiston, M. E. & Haselhuhn, M. P. (2011). 'A Face Only an Investor Could Love: CEOs' Facial Structure Predicts Their Firms' Financial Performance'. *Psychological Science* 22(12): 1478–83.

Wu, C. (2007). 'Queueing Network Modeling of Human Performance and Mental Workload in Perceptual-Motor Tasks', *The University of Michigan* (Dissertation).

Wu, S. (2001). 'Adapting to Heart Conditions: A Test of the Hedonic Treadmill'. *Journal of Health Economics* 20(4): 495–508.

Wyatt, T. D. (2010). *Pheromones and Not (Quite) What You Think.* Poster at International Society for Chemical Ecology Meeting, Tours, France.

Wyatt, T. D. (2011). *Pheromones and Reproduction.* Symposium 1, Society for Reproduction and Fertility.

Yalch, R. & Spangenberg, E. (1990). 'Effects of Store Music on Shopping Behavior'. *Journal of Consumer Marketing* 7(2): 55–63.

Youn, S. & Faber, R. J. (2000). 'Impulse Buying: Its Relation to Personality Traits and Cues'. *Advances in Consumer Research* 27(1): 179–85.

Zadravec, T., Bucik, V. & Sočan, G. (2005). 'The Place of Dysfunctional and Functional Impulsivity in the Personality Structure'. *Horizons of Psychology* 14(2): 39–50.

Zajonc, R. B. (2001). 'Mere Exposure: A Gateway to the Subliminal'. *Current Directions in Psychological Science* 10(6): 224–8.

Zebrowitz, L. A. (1999). *Reading faces: Window to the soul?* Boulder, CO: Westview Press.

Zeitz, K. M., Tan, H. M. & Zeitz, C. J. (2009). 'Crowd Behavior at Mass Gatherings: A Literature Review'. *Prehospital and Disaster Medicine* 24(1): 32–8.

Zimmerman, J. The Effect of Mood on the Preference of Music.

Zuckerman, M. (1991). *Psychobiology of Personality.* Cambridge, England: Cambridge University Press.

Zuckerman, M., Kuhlman, D. M., Thornquist, M. & Kiers, H. (1991). 'Five (or Three) Robust Questionnaire Scale Factors of Personality without Culture'. *Personality and Individual Differences* 12(9): 929–41.

# Index